U.S. Missile Defense Policy and Security in Asia-Pacific Region

by Setsuo Takeda

sanwa

Paul Wolfowitz: Deputy Secretary of Defense in the first and second George W. Bush Administrations, and President of the World Bank

Robert Zoellick: Ambassador for the United States Trade Representative's Office in the first George W. Bush Administration, and Deputy Secretary of State in the second George W. Bush Administration

Richard Armitage: Deputy Secretary of State in the first George W. Bush Administration

Lawrence Lindsey: Assistant to the President for Economic Policy and Director of National Economic Council in the White House in the first George W. Bush Administration, and Governor of Federal Reserve System

Franklin Miller: Special Assistant to the President and Senior Director for Defense Policy and Arms Control in National Security Council in the White House in the first and second George W. Bush Administrations

Brent Scowcroft: Chairman for President's Foreign Intelligence Advisory Board in the first and second George W. Bush Administrations, and National Security Adviser to President Gerald Ford and President George H. Bush

James Schlesinger: Secretary of Energy, Secretary of Defense, and Director of Central Intelligence Agency (CIA) in the Richard Nixon, Gerald Ford and Ronald Reagan Administrations

Harold Brown: Secretary of Defense in the Jimmy Carter Administration, and Supreme Adviser to the Democratic Party

James Woolsey: Director of Central Intelligence Agency (CIA) in the Bill Clinton Administration, and Chairman of the Defense Policy Board in the first and second George W. Bush Administrations

Jeane Kirkpatrick: U.S. Ambassador to the United Nations in the Ronald Reagan Administration, and Senior Member of the Defense Policy Board in the first and second George W. Bush Administrations

Zbigniew Brzezinski: National Security Adviser to President Jimmy Carter, Member of the President's Foreign Intelligence Advisory Board, and Trustee of the Trilateral Commission

James Steinberg: Deputy National Security Adviser to President Bill Clinton , Board of Directors in Pacific Council on International Policy, and Vice President of the Brookings Institution

Jesse Jackson: Civil Rights leader in the U.S., President of Rainbow/Push Coalition, and Democratic Party Candidate in 1988 Presidential election

Richard Lugar: Senator, Chairman of Senate Foreign Relations Committee, and Former Republican Party Presidential Candidate

Chuck Hagel: Senator, and Chairman of International Economic Policy, Export and Trade Promotion Subcommittee in Senate Foreign Relations Committee

Craig Thomas: Senator, and Chairman of East Asian and Pacific Affairs Subcommittee in Senate Foreign Relations Committee

James Inhofe: Senator, Chairman of Senate Environment and Public Works Committee, and Member of Senate Armed Services Committee

Daniel Inouye: Senator, Chairman of Senate Indian Affairs Committee, and Member of Homeland Security Subcommittee in Appropriations Committee.

Lindsey Graham: Senator, and Member of Senate Armed Services Committee

Floyd Spence: Congressman, and Chairman of Armed Services Committee

Henry Hyde: Congressman, and Chairman of International Relations Committee

Curt Weldon: Congressman, Vice Chairman of Armed Services Committee, and Chairman of Tactical Air and Land Forces Subcommittee

Dan Burton: Congressman, Chairman of Government Reform Committee, and Member of International Relations Committee

Jim Turner: Congressman, Ranking Member of Homeland Security Committee, and Member of Armed Services Committee

Ike Skelton: Congressman, and Ranking Member of Armed Services Committee

John Larson: Congressman, Ranking Member of House Administration Committee, and Member of Armed Services Committee

Peter King: Congressman, Chairman of Domestic and International Monetary Policy, Trade and Technology Subcommittee, and Members of International Relations and Homeland Security Committees

Madeleine Bordallo: Congresswoman, and Member of Armed Services Committee

Diane Watson: Congresswoman, and Member of International Relations Committee

Robin Hayes: Congressman, Chairman of Livestock and Horticulture Subcommittee, and Member of Armed Services Committee

Howard McKeon: Congressman, Chairman of 21st Century Competitiveness Subcommittee, and Member of Armed Services Committee

Todd Akin: Congressman, Chairman of Workforce, Empowerment and Government Programs Subcommittee, and Member of Armed Services Committee

John McHugh: Congressman, Chairman of Total Force Subcommittee in Armed Services Committee, and Member of International Relations Committee

Walter Jones: Congressman, and Member of Armed Services Committee

Herbert Bateman: Congressman, and Chairman of Military Readiness Subcommittee in Armed Services Committee

Owen Pickett: Congressman, and Ranking Member of Military Research and Development Subcommittee in Armed Services Committee

John Hostettler: Congressman, Chairman of Immigration, Border Security and Claims Subcommittee in Judiciary Committee, and Member of Armed Services Committee

James Gibbons: Congressman, Vice Chairman of Resources Committee, Chairman of Intelligence and Counterterrorism Subcommittee, and Member of Armed Services Committee

Neil Abercrombie: Congressman, and Ranking Member of Tactical Air and Land forces Subcommittee in Armed Services Committee

Gary Ackerman: Congressman, and Ranking Member of Middle East and Central Asia Subcommittee in International Relations Committee

Howard Berman: Congressman, Ranking Member of Courts, Internet and Intellectual Property Subcommittee in Judiciary Committee, and Member of International Relations Committee

Michael Green: Special Assistant to the President and Senior Director of Asian Affairs in National Security Council in the White House in the first and second George W. Bush Administrations

Richard Fairbanks: U.S. Ambassador-at-large, Chief U.S. Negotiator for the Middle East Peace Process, and Assistant Secretary of State in the Ronald Reagan and George H. Bush Administrations

Michael Armacost: U.S. Ambassador to Japan in the George H. Bush Administration, and President of The Brookings Institution

James Lilley: Assistant Secretary of Defense, U.S. Ambassador to China in the George H. Bush Administration, and U.S. Ambassador to South Korea in the second Ronald Reagan Administration

Charles Pritchard: U.S. Ambassador and Special Envoy for Negotiations with North Korea and U.S. Representative to the Korean Peninsula Energy Development Organization in the second Bill Clinton and first George W. Bush Administrations

Robert Gallucci: U.S. Ambassador and Special Envoy for Negotiations with North Korea and Assistant Secretary of State in the first and the second Bill Clinton Administrations

Richard Solomon: Assistant Secretary of State and U.S. Ambassador to the Philippines in the George H. Bush Administration, and President of United States Institute of Peace

Rust Deming: Assistant Secretary of State for East Asian and Pacific Affairs in the first Bill Clinton Administration

Thomas Pickering: U.S. Ambassador to the United Nations, Undersecretary of State, and U.S. Ambassador to Russia in the first and second Bill Clinton Administrations

Kenneth Bacon: Pentagon Spokesman, Assistant Secretary of Defense for Public Affairs in the first and second Bill Clinton Administrations, and President of Refugees International

John Hamre: Deputy Secretary of Defense in the second Bill Clinton Administration, and President of Center for Strategic and International Studies (CSIS)

Kurt Campbell: Deputy Assistant Secretary of Defense for Asia and the Pacific at the Pentagon, and Senior Vice President of Center for Strategic and International Studies (CSIS)

John Hill: Senior Director for Japan in Asia and Pacific Affairs in the Department of Defense in the first and second George W. Bush Administrations

Patrick Cronin: Consultant for the Department of Defense, Executive Director of the Hills Program on Governance, and Senior Vice President of CSIS

Larry Welch: Air Force General, and President of Institute for Defense Analyses

William Nash: Army General, Military Commander in Bosnia, and Regional UN Administrator in Kosovo

Charles Boyd: Air Force General, and President of Business Executives for National Security

William Kristol: President of Project for the New American Century (PNAC), and Editor of the Weekly Standard

Jessica Mathews: President of Carnegie Endowment for International Peace, and former official in National Security Council and US State Department

Christopher DeMuth: President of American Enterprise Institute (AEI) for Public Policy Research, and former US government official

Alton Frye: President of Council on Foreign Relations, and former US goverment official

Edwin Feulner: President of Heritage Foundation

Christopher Makins: President of The Atlantic Council of the United States

CONTENTS

Foreword 1
Acknowledgements 11
Acronyms and Abbreviations 13

Chapter I Introduction 15
1. Problem Statement 15
2. A Model of U.S. Missile Defense Policymaking 18
3. Missile Defense as a Case Study in Policy Analysis 25
4. Setting the Missile Defense Agenda: The Pentagon Bureaucracy .. 28
5. Missile Defense Policy Adoption and Implementation 31
   A. The President 31
   B. Congress: Motives and Cues 33
6. Policy Impact: Implications for East Asian Security 34
7. Conclusion 36

Chapter II Shaping the Missile Defense Agenda: 1946-2005 39
1. Theory: Missile Defense and Subsystem Politics 39
   A. Iron Triangles 43
   B. Issue Networks 47
   C. Policy Communities 50
2. History and Development of Missile Defense 53
   A. Post-World War II, the ABM Treaty, and Era of Détente, 1946-1983 53

      a. Origins of Missile Defense 53
      b. The ABM Treaty and Era of Détente 56
   B. The "Star Wars" Era, 1983-1993 59
      a. The Strategic Defense Initiative 59
      b. A "New World Order" 61
   C. President Clinton and Theater Missile Defense (TMD), 1993-2001 64
      a. The End of "Star Wars" 64
      b. Assessing the Threat: The Rumsfeld Commission 70
      c. Testing NMD 73
   D. Missile Defense and President George W. Bush, 2001-2005 76
      a. The Missile Defense Agenda 76
      b. Funding the Agenda 77
3. Who Controls Missile Defense Policymaking 78
   A. An Open or a Closed System? 79
   B. The Range of Actors 81
   C. Material Interests 83
   D. Who Controls the Agenda? 84
4. Conclusion 86

Chapter III The President and the Congress : Making Missile Defense Policy Choices 97

1. Introduction 97
2. Theory: Presidential Power and Congressional Motivations ... 101
   A. The President 101
   B. Congress 105
3. Presidential Leadership 108
   A. The Clinton Administration 108
   B. The George W. Bush Administration 118
   C. Evaluation: Clinton v. Bush 123
4. Missile Defense Policy in Congress 125
   A. Advertising: Identifying the Threat 125

B. Position Taking and "Triangulation": The National Missile Defense Act (NMDA) 130
C. Credit Claiming: The 2000 and 2002 Elections 135
D. Congressional Motivations and TMD 139
    a. NMD v. TMD 139
    b. The TMD Improvement Act of 1998 144
5. Conclusion : The changing Balance of Power in U.S. Politics . 150

Chapter IV The Impact of U.S. Missile Defense Policy on East Asian Security 157

1. Introduction 157
2. Theory: The Pluralistic Security Community and the Dangers it Faces 160
3. Impact of U.S. Missile Defense Policy 170
   A. The Korean Peninsula 170
   B. China and Taiwan 182
   C. Chinese Perceptions of U.S. Missile Defense Policy 189
   D. Japan and the U.S.-Japan Security Relationship 192
   E. General Assessment of Missile Defense Policy Impact on East Asian Security 196
4. Indirect Effects on East Asian Security: Missile Defense and the War against Terror 198
   A. The Link between Missile Defense and the War against Terror ... 198
   B. Is Missile Defense the Right Weapon against Terror? 201
5. Conclusion: Caution is the Watchword 206

Chapter V Conclusion 213

1. Review of Findings 213
2. Further Considerations 220
3. Recommendations 223
4. Suggestions for Further Research 226

Index 229

Foreword
by
Robert E. Harkavy

Professor Setsuo Takeda has presented a trenchant critique of emerging American policy towards both its proposed National Missile Defense (NMD) and Theater Missile Defense (TMD) programs, in the latter case as applied to possibilities regarding Japan, South Korea, and Taiwan. In his capacity of long-time student of American politics and foreign policy (and strong friend of the U.S.), Professor Takeda's basic views are that the U.S. should only move cautiously on NMD and TMD, that it should take into account the interests, needs, and views of its allies such as Japan (multilateralism rather than unilateralism as applied to this subject), and that if the U.S. does move ahead, its phased deployments of defensive missiles should be coupled to efforts at arms control and economic measures to encourage détente in East Asia with China and North Korea.

What then is the evolving international political context of these problems as of mid-2004, some three years after the fateful events of 11 September 2001? And, what future scenarios might one speculate upon that might involve actual use of U.S. TMDs in Asia if those defenses were deployed? Alternatively, what might be the implications of not deploying the missiles?

First off, whether or not a now permanent feature of the international security landscape, the U.S. and Russia (Bush and Putin) had, overall, come much closer together, signified by a new agreement on offensive nuclear strategic weapons, Russia's grudging concession of the aborting of the Anti-Ballistic-Missile Treaty (ABM), and Russia's now partial membership in NATO. Generally, the U.S. no longer sees Russia as an enemy or a strategic threat. This U.S.-Russia rapprochement was greatly weakened during the run-up to the U.S. invasion of Iraq in 2003, when Putin sided with France under Chirac and Germany under Schroder. But, the horrible massacre at Beslan in September 2004 by Chechnyan terrorists seemed to drive the U.S. and Russia back closer together in a common front against radical Islamist terrorism.

Relations between the U.S. and China remained problematic and would likely remain so for a long time. There was the looming threat of rival, " hegemonic China", human rights and trade disputes, the South China Sea, the Chinese buildup of strategic nuclear forces, etc., but above all, the nasty and seemingly unresolvable issue of Taiwan. As the U.S. and Russia built their strategic forces down via the START process, China's nuclear forces, probably involving more than 500 nuclear weapons, loomed larger. China's threats to invade Taiwan if the latter did not submit to PRC control, and the U.S. commitment to defend Taiwan against an invasion, seemed to put the two major powers on an inevitable collision course, with some PRC leaders apparently believing that the U.S. would not defend Taiwan if China had the capability to destroy Los Angeles with a nuclear weapon.

In Korea, there were continuing tensions between north and south simultaneously juxtaposed to what may be an inevitable unification of the Korean Peninsula. North Korea was widely assumed to have a few nuclear weapons, based on plutonium production from military reactors, and was also widely believed to be building the capability to produce U-235 from gas centrifuges based on transferred Pakistani technology. Some U.S. intelligence analysts fear North Korea is en route to a stockpile of 40-50 nuclear weapons in the near future. It continued to aid Pakistan, Iran, and some Arab countries with missile technology despite U.S. pressures to the contrary. There was the overarching question of what would be the implications — for the U.S., Japan, China, and Russia — of a united Korea possessing nuclear weapons.

Japan, still with the world's second largest economy, continued to wallow in its seemingly endless economic slump, though there were signs of an impeding recovery. There were questions about the long-term impact of this on Japanese security policies, as Japan faced a future of rising Chinese military and political dominance of Asia. Some reports spoke of a "hollowing out" of the Japanese industrial infrastructure as Japan's own firms moved production to low-cost China, and Japan was less and less spoken of as the leader of an emerging Asian economic bloc. More and more, Japan seemed fated to dependence on the provision of security by the U.S., but some analysts also detected rising sentiment in Japan for more extensive rearmament, perhaps inevitably to include nuclear weapons. It was widely recognized that if it so chose, Japan could rapidly become a major nuclear military power. But, as of 2004, Japan's traditional (since 1945) anti-nuclear pacifism still seemed strongly to hold sway. However, there were also continuing signs, evidenced, for instance, in the Asia Cup football match between China and Japan, of strong Chinese revanchism against Japan because of the experiences during World War II. Similar sentiments are still obvious in Korea. In both cases, this may not bode well for East Asian détente.

Meanwhile, before September 11, but continuing afterwards, the Bush Administration had succeeded in greatly improving relations with India, now to include a defense dimension and greatly increased transfers of military technology. Clearly, the U.S. had come to see a combination of Japan and India as a counterweight to rising Chinese power, modified by the need to retain strong ties to Pakistan in connection with the quest to hunt down Al Qaeda and the background of the Saudi-Pakistan tie. But, in 2004, the Congress Party returned to power in India, perhaps to presage a weaker Indian security relationship with the U.S., and also Israel, while there are some signs of a mellowing of the India-Pakistan confrontation over Kashmir, perhaps to an extent because leaders on both sides are aware of how close their nations have come to the nuclear brink.

Finally, the onward march of developments in weapons

of mass destruction (WMD) in Iran and North Korea put more pressure on the U.S. drive to develop an NMD and TMDs, notwithstanding the assertion by some analysts that the main dangers were insertion of WMDs into the U.S. or the lands of its allies by ships, trucks, or means other than missile delivery. Surely, at any rate, the U.S. public and Congress were now fully beyond the Clinton era of euphoria about the "obsolescence of war", the "transformation of war", and various interpretations of "endism". But, the questions, political and technological, remained — what good is NMD and what good are TMDs?

What scenarios might one envision wherein if a TMD were installed (in Japan, South Korea and Taiwan, or in Japanese territory to cover Taiwan), they might actually come into use? Basically, this involves several scenarios where the U.S. might engage in a conventional "strategic campaign" (as in Desert Storm in 1991 and again in 2001), and where the object of such a campaign threatens to respond with a WMD attack (in a way, analogous to the Iraqi Scud attacks on Israel and Saudi Arabia in 1991, but this time with WMDs).

A Chinese invasion of Taiwan (or, short of that, a blockade by submarines, or intimidation by missile tests, or even actual missile attacks) could result in an American response via an aerial strategic campaign against the Chinese mainland, perhaps focused on the degradation of the Chinese long-range missile capability, still largely based on slow-response time liquid-fueled missiles. To the extent China felt unable or unwilling to threaten the U.S. with nuclear weapons, it might so threaten not only Taiwan, but also Japan, in the latter case assuming that such a threat might deter further U.S. attacks on mainland China (I have elsewhere referred to this as "triangular" or "indirect" second-strike deterrence)[1]. This latter threat could, hypothetically, be reduced by a TMD in Japan; contrariwise, it might provoke a larger Chinese nuclear build-up meant to overwhelm such defenses.

A parallel scenario might involve a U.S. conventional strategic campaign against North Korea if the latter again invaded the ROK (this could be premeditated or, rather, emerge from a crisis). This would be a likely response in view of American reluctance to be dragged into a costly, protracted ground war on the Korean Peninsula. Another scenario could involve U.S. preemption of North Korea's nuclear infrastructure including nuclear storage sites, if the latter were to attempt to "break out" to a larger nuclear arsenal, begging serious questions about whether the U.S.

could locate and destroy Pyongyang's gas centrifuge operations. Still another scenario might involve a U.S. strategic campaign in response to escalated transfers of WMD technology by North Korea to Iran, Pakistan, Syria, etc., if these transfers came to be seen as threatening to the U.S. American TMDs in South Korea and Japan might then be needed to counter a North Korean WMD attack against either or both allies. Again coming into play would be a possible attempt by North Korea to attain "triangular second strike" capability vis-à-vis the U.S. by threatening to attack much nearer and more vulnerable Japan, banking on U.S. reluctance to risk large-scale Japanese casualties. Of course, there is the fundamental question of just how effective a U.S. TMD could be in such circumstances, either in deterrent or defense terms, where there would be no room for error.

It is in the above complex context that professor Takeda provides a Japanese perspective on U.S. national and theater missile defenses. He recognizes that Japan will increasingly live under the shadow of a burgeoning Chinese nuclear capability easily targeted on smaller and more compact Japan; maybe too under the shadow of a much smaller but still threatening North Korean nuclear capability. A U.S. TMD would provide some defense (how comprehensive and when is at issue) against Chinese or North Korean targeting of Japan in lieu of actual targeting of the U.S. — optimally, this might involve a combination of the NAD and NTV systems discussed by Professor Takeda. But, installation of such a defense makes many Japanese nervous and perhaps more inclined to seek security in arms control (maybe trading off TMD for limitations on Chinese or North Korean offensive nuclear capabilities, if that could indeed be brought about, itself a doubtfull proposition) or through economic assistance to its poorer Asian rivals, at least in the case of North Korea. Japan can, of course, try to detach itself completely from Taiwanese and South Korean defense, which might require the elimination of American bases in Japan, including Okinawa. Such are some of the complexities of the missile defense issue.

Robert E. Harkavy
Professor of The Pennsylvania State University
and
Consultant for the Department of Defense (Pentagon)

References
[1] Robert E. Harkavy, "Triangular or Indirect Deterrnce/Compellence." *Comparative Strategy*, Vol. 17, No. 1 (Jan-Mar 1998), pp. 63-81.

Acknowledgements

I appreciate the kind cooperation of many members of Congress (Senators, Congressmen and Congresswomen), high-ranking officials of the federal government, thinktankers and professors in the United States of America. They have provided me with invaluable and useful information, comments, advice, support and help in preparing this analysis.

I am particularly grateful to Dr. Robert E. Harkavy, a consultant for the Department of Defense (Pentagon) and a professor of political science at The Pennsylvania State University, for his insightful comments and advice. I also sincerely appreciate his kindness and cooperation for writing the forward of this book.

A very special debt is owed to two professors who gave me tremendous support and encouragement. Dr. Saburo Sato (Dean) and Dr. Toshiyasu Ishiwatari (Vice Dean) at the College of International Relations of Nihon University kindly encouraged me to write this book. I am deeply grateful to them.

In this way, I have received immeasurable advice, support, assistance, comments and encouragement from those people mentioned above, but they bear no responsibility for the errors and interpretations in this book. The responsibility is solely my own.

Finally I would like to thank Mr. Kou Takahashi for giving me the opportunity to publish this book.

Setsuo Takeda
June 2005

Acronyms and Abbreviations

ABM Treaty	Anti-Ballistic Missile Treaty of 1972
ADI	Air Defense initiative
AEV	Airborne Early Warning
ASDF	Air Self-Defense Force
ASEAN	Association of South-East Asian Nations
ASM	Air to Surface Missile
ASW	Anti-Submarine Warfare
AWACS	Airborne Warning and Control System
BIR	Bureau of Intelligence and Research
BMDO	Ballistic Missile Defense Organization
BUR	Bottom Up Review
CIA	Central Intelligence Agency
CTBT	Comprehensive Test Ban Treaty
DIA	Defense Intelligence Agency
DoD	Department of Defense
DoE	Department of Energy
DoS	Department of State
DPAC	Defense Policy Advisory Committee
FAS	Federation of American Scientists
GSDF	Ground Self-Defense Force
IAEA	International Atomic Energy Agency
ICBM	Inter-Continental Ballistic Missile
INF	Intermediate-Range Nuclear Forces Treaty
IRBM	Intermediate-Range Ballistic Missile
JDA	Japan Defense Agency

JSDF	Japan Self-Defense Forces
LSI	Lead System Integrator
MAD	Mutual Assured Destruction
MD	Missile Defense
MDA	Missile Defense Agency
MEADS	Medium Extended Area Defense System
MOU	Memorandum of Understanding
MRBM	Medium-Range Ballistic Missile
MSDF	Maritime Self-Defense Force
MTCR	Missile Technology Control Regime
NAD	Navy Area-wide Defense
NASA	National Aeronautics and Space Administration
NATO	North Atlantic Treaty Organization
NMD	National Missile Defense
NMDA	National Missile Defense Act
NPT	Non-Proliferation Treaty
NSC	National Security Council
NTWD	Navy Theater-Wide Defense
PAC	Patriot Advanced Capability
SALT	Strategic Arms Limitation Talks
SAM	Surface to Air Missile
SDIO	Strategic Defense Initiative Organization
SDF	Self-Defense Forces
SDI	Strategic Defense Initiative
SLBM	Submarine-Launched Ballistic Missile
SRBM	Short-Range Ballistic Missile
SSM	Surface to Surface Missile
START	Strategic Arms Reduction Talks
THAAD	Theater High-Altitude Area Defense
TMD	Theater Missile Defense
UN	United Nations
USSR	Union of Soviet Socialist Republics
WMD	Weapons of Mass Destruction

Chapter I

Introduction

1. Problem Statement

On the morning of September 11, 2001, the members of the House Defense Appropriations Subcommittee were meeting in the US Capitol to begin work on the annual defense authorization bill. The agenda included an $8.3 billion request from President George W. Bush to fund an overall missile defense program, and in particular a national missile defense (NMD) system to protect the US from attacks by "rogue" nations. The meeting was suddenly canceled and the building evacuated when the members learned that terrorists had seized two jetliners in midair and crashed them into the World Trade Center in New York City. A third plane, similarly hijacked by terrorists, had plunged into one side of the Pentagon and a fourth went down in a Pennsylvania field, apparently en route to a collision with the US Capitol Building.[1] In a few hours, the

American defense agenda had changed irrevocably. Eventually, American leaders and citizens alike would have to face the question of whether missile defense had continued relevance in an age of terrorism.

Three months later, on December 20, 2001, both houses of Congress gave their answer when they agreed to a defense appropriations bill, which was later signed by President Bush. In the interim, disputes arose between Congress and the president over leaked information to the press, executive secrecy, and other specific provisions that would benefit one specific defense contractor, the Boeing Corporation.[2] The final bill, however, included $7.8 billion for missile defense, a 50 percent increase over President Clinton's final budget in this area, but $500 million less than President Bush had requested.[3]

Clearly, both President Bush and the Congress agreed that US security continued to require a missile defense system featuring NMD despite the fact that the most violent attack upon the country's civilian population from a foreign enemy in its entire history had just occurred, and this attack did not involve offensive missiles launched from any foreign military base. Instead, this attack was the product of infiltration, stealth, ruthlessness and suicidal boldness on the part of a small number of avowed terrorists who used simple knives to commandeer commercial aircraft and turn them into weapons of mass destruction by crashing them into buildings. The only defense against such an attack would have included, first, the ability to gather accurate information from surveillance of individuals and groups capable of such attacks, and secondly, the capacity to communicate this information effectively and efficiently to those who are capable of taking action to thwart the attack. Information is as important as weaponry as a component of any set of countermeasures against terrorism.

Why, then, does missile defense continue to demand such enormous support within the US government? The answer, I think, is partly connected to a genuine military strategy of deterrence which its advocates believe will strengthen the interests of both the US and its allies to a greater degree than any other such strategy given the realities of the current international system. Another part of the answer, I think, is related to the dynamics of current American politics, the interactions of the President and the Congress in the making of security policy, the competition between the major parties for control of the political branches of the government, and the interactions of organized interests in the American policymaking process.

What is the impact of US missile defense policy on its allies and on the international system? The answer to this question depends greatly on the changing dynamic of the international system. Many in the US believe that the willingness of the administration of President Ronald Reagan to initiate research and development on missile defense in 1983 forced the Soviet Union into a downward slide that led to the collapse of the regime. Today, some believe that the same strategy could be used to bring down "rogue" regimes in such places as North Korea, and perhaps even China. The danger of such a strategy, however, is that missile defense programs, like NMD, may cause potential US adversaries, especially in East Asia, to invest more heavily in offensive weapons and produce a dangerous arms race threatening stability in the region.

Is this danger acknowledged by decision-makers in the US? In the next three chapters I will explain why I think that it is not recognized to the degree that it should be, and I will offer a few rough suggestions as to the range of prudent responses available to US allies in East Asia. I will also offer suggstions as to how the process of making missile defense policy in the US might be improved.

2. A Model of U.S. Missile Defense Policymaking

Missile defense has had a long history in American defense policy dating back to the earliest days of the Cold War and the arms race between the US and the Soviet

Union. Even after the collapse of the Soviet government and the subsequent calls within the US for reductions in military spending, however, missile defense remained high on the agenda of American defense policy makers. The rationale for this was essentially that the post-Cold War world remained a dangerous place, one in which "rogue states" could obtain high technology weapons of mass destruction despite their relative social and economic backwardness. These "rogue states"— usually said to include North Korea, Iraq, Iran and Libya — were the new threat that justified further development of missile defense systems.

The Persian Gulf War of 1991 between Iraq and a military coalition led by the US demonstrated the validity of this argument. In that conflict, Iraqi missiles were used to attack coalition forces and the Israeli city of Tel Aviv. Although both the Iraqi offensive missiles and the coalition's missile defenses were almost equally ineffective, the lesson drawn from the experience was that missile defense systems, including missile defense, had a continuing usefulness in the post-Cold War era. Missile defense remained a campaign issue in American national elections. The funding of missile defense programs continues to be debated in Congress, the Pentagon, and the White House. The political parties were divided over the issue.

While these events were unfolding in American national

politics, however, a new threat was beginning to emerge into greater prominence. After the 1993 attempt to bomb the World Trade Center in New York City, policymakers and the public in general were forced to take seriously the threat of non-state terrorist organizations, such as Al-Qaeda, using relatively primitive, homemade weapons to attack symbolic targets throughout the developed world. Against such weapons, missile defense was of no value. Although high-tech missiles with precision accuracy were useful in counter-attacking terrorist bases hidden in remote locations, such as the mountains of Afghanistan, they could not prevent or even deter such attacks. Nevertheless, American policy continued to stress the development of a variety of missile defense systems throughout the 1990's with an increasing sense of urgency and with higher levels of funding for missile defense research and development.

Missile defense became an issue in the presidential election campaign of 2000 when George W. Bush promised that, if elected, he would deploy NMD within four years. In 2001, after his victory, President Bush moved quickly to nearly double missile defense funding and to nullify the ABM Treaty with Russia that had stood as a barrier to missile defense deployment for nearly thirty years. During the early months of his administration, President Bush focused many of his efforts on negotiating the provisions of the

Figure 1.1 A Model of Missile Defense Policy Making

1. Agenda Setting and Policy Formulation	2. Policy Adoption	3. Policy Implementation	4. Impact on the International System
Institutions:			
President and Department of Defense ▶	President and Congress ▶	Missile Defense Agency ▶	Nation States
Actions:			
Integration with over all strategy; Research and development; Initial budget request.	Final decisions on strategic implications; Final budget appropriations.	Selection of contractors; Procurement and deployment.	Regime change; Arms build-up; Increase in international tensions.

Source: Created by the author

missile defense programs with Congress. He also tried to assure US allies in Europe and Asia that US missile defense systems would enhance their security as well as that of the US. As I have already mentioned, during the terrorist attack on September 11 of that year, missile defense was a prominent item of discussion in several key committees of Congress.

This emphasis on missile defense may have blinded American leaders to the more salient threat of international terrorism, but this question is beyond the scope of this research. I have chosen to analyze missile defense policy employing the model shown in Figure 1.1. This model also serves to illustrate the plan of this research.

Figure 1.1 divides the subject matter of Chapters Two, Three and Four into four areas of policy making. Chapter Two will cover the agenda setting and policy formulation stages, which, in the case of missile defense, are dominated by the president as commander-in-chief and by the bureaucracy in the Department of Defense. Actions taken by these institutions at these stages include the integration of missile defense with overall security strategy, research and development of missile defense systems and the creation of initial budget requests. These instituions and actions are discussed in Chapter Two.

Chapter Three moves my discussion to the policy adoption stage, which engages the president and Congress in the final decisions on the role of missile defense in overall strategy and in the final budget appropriations for missile defense. These actions lead to the implementation of policy by the Missile Defense Agency, which was originally known as the Strategic Defense Initiative Organization (SDIO) and later as the Ballistic Missile Defense Organization (BMDO). At this stage, private contractors are chosen, the systems are tested which lead to final procurement and deployment.

Finally, in Chapter Four, I explore the impact of these

actions on the international system, specifically in the various hot spots of East Asia. Here I expect that US missile defense policy might be to effect regime change in those states; but it might also be a build-up of weapons systems and an increase in international tensions.

Seen in this way, missile defense becomes a case study showing how numerous complex interactions among elements of American domestic politics shape US military policy. Current high levels of partisan competition in American national politics affect the motivations and behaviors of presidents, members of Congress, military analysts in the Pentagon and interested individuals and groups outside of government. In such a climate, any issue, any decision, any policy — even one like missile defense, which ordinarily would lie outside the range of general public discussion — can significantly alter power arrangements within American political institutions. This reality affects all decisions, but may also profoundly affect the choices US makes in defending itself.

I also want to look at the impact of missile defense on international relations in East Asia in order to demonstrate that these domestic political processes have important, if indirect, effects on the international system. In East Asia, I believe that the deployment of theater missile defenses (TMD) will affect at least two security issues: 1.) nuclear proliferation on the Korean Peninsula, and 2.) military dangers and tensions in the Straits of Taiwan. Partners of the US in the international system may consider missile defense to be an umbrella, which could shelter them and deter others from aggressive action. In a rainstorm, however, the one who holds the umbrella is the one most likely to remain dry. For the future, the others may wish to have umbrellas of their own, which would come, of course, in the form of some considerable increase in military preparedness on their part — preparedness which will be developed in a way that is more independent of US influence than in the past.

To put my case concisely, I think that missile defense is an important link between domestic political processes in the US and events in the international system that affect the security of East Asia. I see the process of making missile defense policy as the product of a dynamic system whose actions are shaped by American domestic politics and whose impacts are global. The intensely partisan American political environment today has made missile defense an issue that divides the major political parties. Although relatively few voters see missile defense policy by itself as a salient issue in political campaigns, the larger question of national security may make missile defense a critical question capable of tipping the balance of power between the major parties in national elections.

Partisanship over missile defense has affected a variety of institutional relationships. The Department of Defense has felt increasing pressure to dramatize the need for missile defense and to demonstrate its value not only militarily but also as a diplomatic bargaining chip. Recent American presidents have bargained with foreign leaders on the one hand and with congressional leaders on the other to produce packages of military expenditures, which include missile defense and satisfy relevant political constituencies both at home and abroad. Meanwhile, members of Congress have pursued electoral success by balancing the competing demands of interest groups at home and bureaucrats in Washington. These elements of the domestic political system will be taken up in turn in each of the next three chapters.

3. Missile Defense as a Case Study in Policy Analysis

I consider missile defense to be an example of the public policy output of a complex system. I emphasize that public policy is a course of action directed toward some goal. To put it more formally, public policy is

"..... a proposed course of action of a person, group, or government within a given environment providing obstacles and opportunities which the policy was proposed to utilize and overcome in an effort to reach a goal or realize an objective or a purpose."[4]

Any policy, such as missile defense, as a purposive course of action, is the result of the workings of a complex system of the sort described by Easton as comprising identifiable and interrelated institutions and processes.[5] In Chapters Two, Three and Four, in order to effectively address the question of how governmental institutions interact to produce public policy, I have adopted a strategy which involves a blend of systems analysis and institutionalism. This allows us to look into the "black box" to see how the system itself operates within institutions to produce a policy such as missile defense.[6] I will be looking, for example, at how President Reagan preserved missile defense against its political opponents through the creation of an agency within the Pentagon with the specific function of developing missile defense policy. I will also look at how presidents have negotiated with Congress for missile defense budgets that reflected their own preferences with regard to missile defense and also guarded their own larger political advantages. The combining of systems analysis and institutionalism makes this possible.

I consider the making of missile defense policy to be a

case study of how domestic politics is linked to international consequences. In this sense I regard missile defense as an "intermestic" policy, defined by Cochrane and Malone as one which involves "a mixture of international and domestic policies."[7] On the one hand, the proponents of missile defense base their arguments on the necessity of the US to counter threats from rogue or new nuclear weapon states. On the other hand, however, members of Congress who support missile defense often refer to the number of jobs created in their states and districts by defense industries related to missile defense rather than to any perceived increase in national security accomplished by missile defense. This may seem like "politics as usual" since leaders often support the policies they prefer by reference to short-term material benefits to their constituents. In this case, however, there may be unseen long-term risks at the point where missile defense connects to the international system, which few leaders appear ready to take into account.

On the international side of missile defense policy, missile defense is, in theoretical terms, an "actor disturbance" which affects the "pattern of international outcomes."[8] In the international system, these outcomes are also affected by "regulators," which take the form of international organizations, such as the United Nations (UN), or alliances, such as the one that exists under the US-Japan Security

Treaty. In addition, there is always an array of "environmental constraints," such as geographical or socioeconomic factors, which limit the range of behaviors available to a nation state.[9] This would suggest that, as a research strategy, missile defense should be treated as an independent variable having a causal affect on the behavior of international actors. What I do instead is to treat missile defense as an intervening variable, which has been produced by a complex set of factors at the domestic level, but which, in turn, also produces a number of consequences at the international level. This allows us to focus attention on the institutional and systemic explanation for missile defense as an output of the domestic US system without losing sight of the fact that missile defense is also a powerful input to the international system. This permits a more comprehensive picture of missile defense in both contexts.

4. Setting the Missile Defense Agenda: The Pentagon Bureaucracy

Chapter Two is a historical account of the evolution of US missile defense policy. Its purpose is to explore the question of who sets the missile defense policy agenda. In this chapter, I will see that from the beginning missile defense policy has been the output of a domestic political system.

America's first version of missile defense was the Nike-Zeus system developed by the Army in the late 1950's and early 1960's to counter the Soviet Union's first intercontinental ballistic missiles (ICBM's). It had limited capabilities, and acted more as a symbol for defense than an actual defense against nuclear-armed missiles. The technological problems involved in hitting or in any other way defending against such missiles were not solved, and eventually, Nike-Zeus was scrapped. In this period in the development of missile defense policy, offensive nuclear weapons programs held a higher place than missile defense on the national security agenda.

Twenty years later, in 1983, President Ronald Reagan issued a surprise announcement that he was ordering a comprehensive and intensive research and development program, which he said would render nuclear weapons impotent and obsolete. This led to President Reagan's Strategic Defense Initiative (SDI), also known as "Star Wars," the precursor to today's missile defense systems. In 1984, President Reagan created the Strategic Defense Initiative Organization in the Pentagon to skirt ordinary bureaucratic budgeting and decision-making structures. From these beginnings, missile defense became the largest research and development project in American history.[10]

My examination of bureaucratic decision-making and

the role of the Pentagon with regard to missile defense will employ an account of the history of the development of missile defense policy as told by the Missile Defense Agency (MDA) and by journalists and analysts who have many of the events firsthand. My analysis will attempt to test the "iron triangle" approach to missile defense policy, which argues that missile defense is similar to other products of the triadic relationship that often exists among congressional subcommittees, bureaucratic agencies and the interest groups who benefit from their decisions.[11] Indeed, missile defense policy has been at times subject to "capture" by the defense contractors who perform much of the research, development and production of the military hardware.[12]

Two alternatives to the iron triangle are the "issue network" approach in which decisions are much more open and where there is much more opposition to the policy preferences of individual actors, and the "policy community" approach, which occupies a middle ground between iron triangles and issue networks. My account of history will show, consistent with the account of Bradley Graham that no one of these approaches can adequately account for missile defense policy.[13] Key presidential decisions have helped to insulate missile defense from political opposition over time since President Reagan created the Strategic Defense

Initiative Organization (SDIO) within the Pentagon in 1984. But these decisions did not prevent open conflict within Congress and between Congress and the president over missile defense.

5. Missile Defense Policy Adoption and Implementation
A. The President

This brings us to Chapter Three and another stage of missile defense policymaking. From President Reagan's 1983 speech launching SDI to the present, presidents have found it necessary to use all of the tools of presidential power in order to achieve their preferences on missile defense. The theme of the first part of Chapter 3 examining the role of the president in making NMD policy underlines the famous judgment of President Harry Truman as recorded in Richard E. Neustadt's landmark study of presidential power: "the power of the president is the power to persuade."[14] I will focus my attention primarily on Presidents Bill Clinton and George W. Bush. These men have found it difficult to command others to do as they wished regarding missile defense; but instead have had to bargain and negotiate with congressional leaders, Pentagon planners, and key figures within their own administrations.

My method in Chapter 3 is to analyze missile defense in

much the same way as others have analyzed presidential decisions regarding the deployment of US troops to foreign military operations. Such operations range from the wars in Korea and Vietnam to the invasions of Grenada or Panama. I think that troop deployment decisions are analogous to missile defense because, like the arguments favoring missile defense, arguments favoring foreign troop deployments center on similar national security concerns. Cohen and Nice have reviewed the literature in this area and have identified two propositions to explain why presidents commit US troops.[15] The first is called "the external threat argument," which suggests that presidents respond to the actions of foreign nations that have threatened the US, its citizens, or its property. The second is referred to as the "domestic or internal politics model," which holds that domestic political pressures create the necessary conditions leading to presidential decisions to commit US troops.

I think the same models can help us understand the presidential decisions of Clinton and Bush regarding missile defense. In my analysis I will compare the overall annual budget requests of these presidents for missile defense, of which missile defense is only a part, with the annual appropriations by Congress for missile defense. Combined with anecdotal evidence, this will permit us to make some tentative judgments about the success of each of these presidents in making missile defense policy and achieving their goals regarding missile defense.

B. Congress: Motives and Cues

In Chapter Three my analysis of the effect of the US Congress on missile defense policy focuses on the motivations of the members and the ways in which information affects their voting. I agree with Mayhew that the members of Congress "are interested in getting reelected," and that fact influences the kinds of activities they find it "electorally useful to engage in."[16] These activities include claiming credit for policies that positively affect constituents and taking positions on issues that are of interest to voters in their home states and districts. I also agree with Matthew and Stimson, who argue that on many complex issues members who occupy key positions and possess the greatest amount of information provide cues to others as to how they should vote on legislation regarding that issue. In Chapter Three I will show that although missile defense is not a salient issue for many members of Congress in either house, the desire to be reelected forces them to take a position on missile defense. My analysis of the debate and roll call votes on the National Missile Defense Act (NMDA) of 1999 will also show that the kinds of information relied upon by members of Congress in deciding their votes on that bill were almost exclusively relevant to their domestic political concerns. The potential international impact of missile defense carried little weight at that time, nor were there many members of Congress prepared to introduce such information into the debate.

6. Policy Impact: Implications for East Asian Security

Having explored the domestic political institutions that have adopted and implemented missile defense, I next turn my attention to the impact of missile defense on East Asian security issues in Chapter Four. In this chapter, I go beyond the domestic political system of the US to the international system, what Hoffman described as "a pattern of relationships among the basic units of world politics."[17] In this system, as in the domestic political system, the level of conflict is so delicately balanced that quite literally every decision or policy adverse to the interest of maintaining the balance may cause the balance to tip dangerously.

Missile defense is a policy with profound implications. My analysis on this point is consistent with the arguments of Richard Butler, the former head of the United Nations Special Commission to Disarm Iraq from 1997-1999, who argues that for the US to expect that other nuclear powers will accept the idea that missile defense is a strictly defensive weapons system is too much to ask. Butler points out that other nuclear powers, and by extension those countries that may acquire nuclear capability, will believe, rightly or wrongly, that the "unstated motive for missile defense is, in fact, to achieve US dominance of space."[18] This is likely to spark a dangerous escalation in weapons competition.

My analysis will focus on events surrounding the most salient current conflicts in East Asia. This includes : 1) the crisis surrounding nuclear proliferation on the Korean Peninsula, 2) the tensions between China and Taiwan, and 3) the spread of terrorism in East Asia. I think the effect of missile defense is to heighten tensions in both of the first two areas of conflict, making peaceful resolution of the crises there far more difficult. At the same time, I think missile defense has little direct effect on the third area other than to divert US energy away form addressing the problem.

I will also consider what courses of action are open to Japan and other US allies in East Asia that are directly affected by these crises. The question I explore in this part of the chapter is, generally, how will these nations judge the efficacy of American decision making in current international crises? Both missile defense deployment and the war in Iraq, which has successfully removed Saddam Hussein from power in that country, have been offered as parts of a program designed to create a safer and more stable international system. Similarly, confrontational tactics by US negotiators in six-nation talks with North Korea have been undertaken with the goal of disarming the North Korea and removing the nuclear threat from the Korean Peninsula. Have any of these policies accomplished their purposes? What have been the results so far? How do US allies in East Asia evaluate US leadership in these crises? We must expect that any nation that finds US leadership to be problematic will seek a more independent path in future crisis situations.

7. Conclusion

In Chapter Five, the conclusion, I will bring together the various conclusions I have reached in order to show the overall implications of my research. I believe that one significance of my study lies in the use of systems theory and institutionalism to examine both the effects of domestic political processes on the making of national security policy and the impact of the resulting policy in the international system. I think this is an unusual but valuable form of policy analysis.

More important, of course, is the recognition that the impact of these processes on the international system is quite significant. Domestic political processes have a great deal to do with agenda setting, policy formulation and adoption, and the implementation of a security policy with powerful international ramifications. There is danger that such policies will be ill-suited to the real security problems that face American allies, especially in East Asia. I think this is an important defect in US foreign policymaking, which requires consideration by foreign leaders and political scientists.

Finally, I suggest that missile defense policy fits a pattern that includes American confrontational tactics in both the North Korean crisis and the continuing tensions in the Straits of Taiwan as well as the military actions in Iraq. For the US leadership, all of these initiatives have served to strengthen the hand of the Bush Administration at home among its base voting constituency while simultaneously destabilizing the international system and creating real dangers for other nations around the world, including US allies. Many nations now back away from support for US policies, and as a result the international system is now adrift without leadership in which most nations can have confidence. A shift affecting future leadership may now be under way in the pluralistic security community, which includes Europe and Japan. The effect of this shift may be to reduce dependence on US leadership and lead to greater independence on the part of other developed nations.

Notes
1 "Anti-Terror Funds Slow Defense Bill," 2001 *Congressional Quarterly Almanac*, pp. 2-13.
2 Ibid.
3 Ibid.
4 Carl J. Friedrich, *Man and His Government* (New York: McGraw-Hill, 1963), p. 79.
5 David Easton, "An Approach to the Analysis of Political Systems," *World Politics*, IX (April 1957), pp. 383-400.
6 James E. Anderson, *Public Policymaking*, 3d ed. (New York: Houghton Mifflin Co., 1997), p. 35.
7 Charles L. Cochrane and Eloise F. Malone, *Public Policy: Perspectives and Choices* (New York: McGraw-Hill, 1995), p. 410.
8 Richard N. Rosecrance, *Action and Reaction in World Politics: International Systems in Perspective* (Boston: Little Brown, 1963), p. 229.
9 Kenneth N. Waltz, *Theory of International Politics* (New York: Random House 1978), pp. 41-43.
10 Bradley Graham, *Hit to Kill* (Cambridge, MA: Perseus Books, 2001), p. 11.
11 Robert L. Lineberry, *American Public Policy: What Government Does and What Difference It Makes* (New York: Harper & Row, 1977), p. 55.
12 Theodore J. Lowi, *The End of Liberalism: Ideology, Policy, and the Crisis of Public Authority* (New York: Norton, 1969), Chapter 3.
13 Bradley Graham, op. cit, p. 383.
14 Richard E. Neustadt, *Presidential Power* (New York: John Wiley and Sons, 1960), p.41.
15 Jeffrey Cohen and David Nice, *The Presidency* (New York: McGraw-Hill, 2003), pp. 444-445.
16 David Mayhew, *Congress : The Electoral Connection* (New Haven, CN : Yale University Press, 1974), p.5.
17 Stanley Hoffman, "International Systems and International Law," in Hoffman, *The State of War: Essays on the Theory and Practice of International Politics* (New York: Praeger, 1965), p.160.
18 Richard Butler, *Fatal Choice: Nuclear Weapons and the Illusion of Missile Defense* (Boulder, Colorado: Westview Press, 2001), p. 107.

Chapter II

Shaping the Missile Defense Agenda: 1946-2005

1. Theory: Missile Defense and Subsystem Politics

The central question of this chapter is: what political forces have controlled the missile defense agenda in the US? To put it more specifically, is this agenda widely debated and governed by popular will, or is it kept out of the public eye and controlled by a few? I will use historical data to determine how the modern US policy of national missile defense (NMD), an elaborate system of land-based defensive missiles, emerged from the tactical problem of how to defend the US against potential attack from hostile enemy missiles. As I will show, this evolving military policy has been the product of many complex institutional interactions within the US government over a period of many years.

Control of the policy agenda in any area is truly an age-old problem in terms of both practical and theoretical politics. During the founding period in American history, James Madison urged his generation of Americans to adopt the Constitution of 1787 by arguing that in a large and diverse nation governed by a system of separated and expressed powers, the ability of any single group, large or small, to dominate the government at the expense of the public good would be checked by various institutional arrangements. A small group would be controlled by the "republican principle" by which the majority is able "to defeat its sinister views by regular vote."[1] The majority, for its part, would be held in check by the "variety of parties and interests" found in a large and diverse republic.[2] The modern state, however, often appears at odds with this pluralistic view. More than a century-and-a-quarter later, Vilfredo Pareto observed that over time, power in any organization tends to become concentrated in fewer and fewer hands and that "whether universal suffrage prevails or not, it is always an oligarchy that governs."[3] More recently still, Max Weber noticed the difficulty in a modern democratic state to limit the power of the civil servant while strengthening the power of the voter. "The most decisive thing," he noted, "is the leveling of the governed in opposition to the ruling and bureaucratically articulated group, which in turn may occupy a quite autocratic position, both in fact and in form."[4] These rival points of view lie behind virtually every debate about how policy is made in any democratic government and provide a framework for this narrow range study of American missile defense policy.

In this chapter I will explore the question of how US missile defense policy is made through a brief examination of the evolution of that policy from the end of World War II through the administration of President George W. Bush. I divide this historical analysis into four parts. The first takes us from the closing days of World War II when the British began considering a defense against German V-2 rockets to the height of the Cold War arms race between the US and the USSR when the Anti-Ballistic Missile Treaty (ABMT) between the two countries sought to limit such weapons. The second part discusses the era of "Star Wars", or more correctly the era of the Strategic Defense Initiative (SDI), which was begun by President Ronald Reagan in 1984 and continued until the election of President Bill Clinton in 1992. The Clinton Administration represents part three of this history, a period in which the administration placed an emphasis on theater missile defense (TMD) at the expense of NMD. During this period, the president and Congress debated and negotiated vigorously over missile defense budgets. Part four covers the presidency of George W. Bush, and is marked by a much closer consensus between president and Congress on missile defense policy.

As I will show, historical analysis reveals significant changes over time in US missile defense policy. My purpose is to test my theory that these changes have occurred as a result of changes in the relationships among policy making institutions within government and groups outside government that seek to influence missile defense policy. In this effort I am aided by three models, which have been developed by political scientists in the field of public policy. The first of these is the "iron triangle" model, originally outlined by J. Leiper Freeman.[5] This model emphasizes tight control over policy making by congressional committees, executive agencies and the interest groups that benefit from their actions. The second model is the "issue network," developed by High Heclo,[6] which describes a far more open policy making process in which many actors, both inside and outside of government, compete with each other to influence the direction of policy. The third model is the "Policy communities", created by James B. Anderson,[7] It lies between these two extremes with less institutional control and more openness than the iron triangle, but more control and less openness than the issue network. I will employ a combination of journalistic reports, government documents and press releases to determine which of these three models over time best explains missile defense policymaking.

A. Iron Triangles

A basic assumption underlying this study of missile defense policy is that any particular public policy is the product of some form of subsystem politics. By this I mean processes through which decisions are made by subunits of larger institutions and processes. Congressional committees and subcommittees, for example, are subunits of Congress. Bureaucratic agencies are subunits of the executive branch, and so on. They carry on their work of reviewing and amending proposed legislation largely outside the glare of mass public attention. As committees and subcommittees, they are not directly accountable to the electorate. Their members are accountable, of course, but their activities as members are rarely a matter of scrutiny in national election campaigns. Most often, the voters are satisfied if their representative has power on a committee and uses that power to benefit his or her constituency.

One approach to the analysis of subsystem politics, therefore, leaves out the role of electoral accountability in explaining public policy, but rather examines triadic relationships among the agencies of the federal bureaucracy, congressional committees and subcommittees and the various client groups that benefit from government spending.

These relationships have been called "iron triangles," "cozy triangles," "sub-governments," "triple alliances," and other even less flattering names. Whatever term is used, it is meant to imply that agreements are struck within these relationships whereby a group operating in the environment of the political system obtains a benefit from government through the combined efforts of the congressional committees that hold the purse strings and the agencies of the bureaucracy that implement public policy in the area of the group's activity. (I will use the term "iron triangle" because it appears to be the one in most common usage.) As Figure 2.1 on the next page illustrates, the goal of these relationships is the mutual material benefit of all concerned.

Figure 2.1 depicts a hypothetical iron triangle showing the relationships among the key actors affecting missile defense policy. In an iron triangle, each actor desires an array of resources, usually involving money and information, in order to perform a specific function, and each one has its desires met by the other actors in the relationship. Each actor also possesses an array of resources desired by the others. Members of congressional committees require information about the potential effectiveness of public policy proposals as well as the possible impact of policy on their constituents. The agencies of the bureaucracy seek to jus-

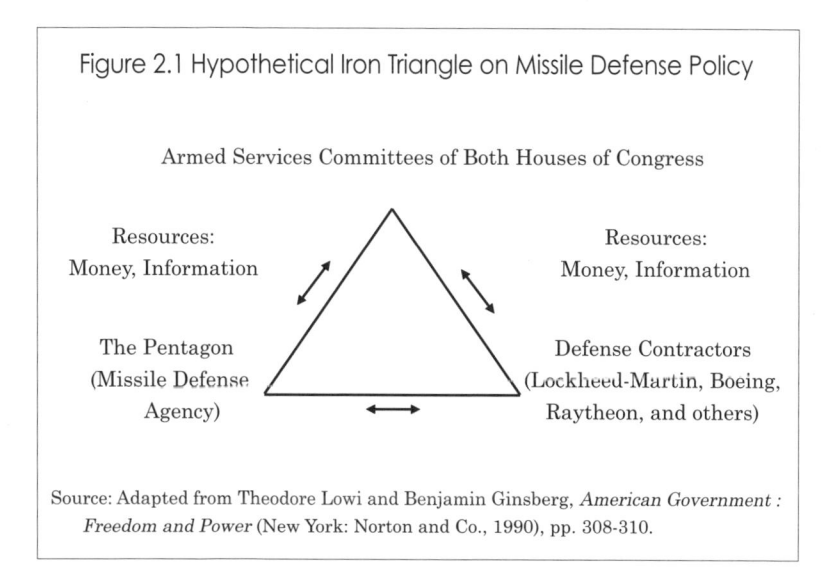

> **Figure 2.1 Hypothetical Iron Triangle on Missile Defense Policy**
>
> Armed Services Committees of Both Houses of Congress
>
> Resources: Money, Information
>
> Resources: Money, Information
>
> The Pentagon (Missile Defense Agency)
>
> Defense Contractors (Lockheed-Martin, Boeing, Raytheon, and others)
>
> Source: Adapted from Theodore Lowi and Benjamin Ginsberg, *American Government: Freedom and Power* (New York: Norton and Co., 1990), pp. 308-310.

tify their budget requests by supplying the committees with the desired information. Client groups lobby the members of Congress providing support for the budget request in the expectation that government contracts or other benefits will result when the budget is approved.

In this case, the Armed Services Committees of both houses of Congress hold the power to approve and set the level of funding for any missile defense policy desired by the Missile Defense Agency (MDA). The agency provides information to Congress regarding the need for, likely effectiveness of, and cost of missile defense. The defense contractors compose the client group, which supports the MDA budget request by further informing the members of the committees on the technical feasibility of various missile defense systems.

The principal characteristic of this three-sided relationship, according to the "iron triangle" approach, is that it is a closed system. The relationships are stable and ongoing. All members of the relationship have a direct material interest in the policy matters being treated. Although budgeting and other decisions regarding any policy are made in public—as all such decisions must be—they are done quietly and out of the glare of public attention. Thus, in an iron triangle, there is little public controversy over policy, and little disagreement among the members of the iron triangle. Operating in this manner, such iron triangle relationships may wield enormous power largely unchecked by ordinary political processes.

This model might aptly describe the development of missile defense policy. MDA works on a budget appropriated by Congress and largely determined by the Armed Services Committees of the two houses. The primary beneficiaries of these budget allocations, aside from the agency itself and Department of Defense, are the principal contractors who build missile defense systems, namely the Boeing Corporation, Raytheon Corporation and Lockheed-Martin Corporation as well as a variety of other contractors and subcontractors who provide goods and services to the overall effort. When these corporations lobby Congress in support of the missile defense budget, they behave in a manner that is consistent with the predictions of the iron triangle model.

The MDA budget, as part of the overall defense budget, is developed within the agency and approved by the Department of Defense and the White House Office of Management and Budget following well-established procedures. It is proposed in Congress and referred to appropriate committees. Defense contractors lobby heavily for approval of the budget. These activities are usually highly effective in gaining congressional approval of the MDA budget, which in turn works to the mutual benefit of each member of the triad. MDA obtains the funds necessary to continue research, development and procurement of missile defense systems; the missile defense industry obtains lucrative government contracts from MDA; and the members of Congress claim credit for improving the defense of the nation while some receive the added bonus of higher employment and other financial benefits for their districts and states through the economic activity thus generated.

B. Issue Networks

The iron triangle model best describes those policymaking activities where there is little public discussion or media attention. The difficulty in applying this model to US missile defense policy making, however, stems from the inevitable public controversy surrounding its technological feasibility, its cost, its effectiveness, and its potential negative consequences. Missile defense may occasionally gain a relatively high level of salience as a national political issue as it did for a time in the 1996 and 2000 presidential elections. When this happens, it is impossible to avoid widespread public attention or to escape pro and con debates over the issue.

This exposes the most obvious weakness of the iron triangle model. In areas of policy making where debate is raging, this model is vulnerable to Heclo's criticism that it is "not so much wrong as it is tragically incomplete."[8] Missile defense is often not the kind of a narrow public program in which a small set of actors with direct material interests at stake can control the substance of policy without drawing significant attention from other actors who oppose their agenda.

A more useful way to develop an explanatory model, therefore, may be to treat US missile defense policy as the product of an issue network," as described in Figure 2.2 on the next page. As the figure shows, the issue network on missile defense policy extends well beyond the armed services committees of Congress, the Missile Defense Agency

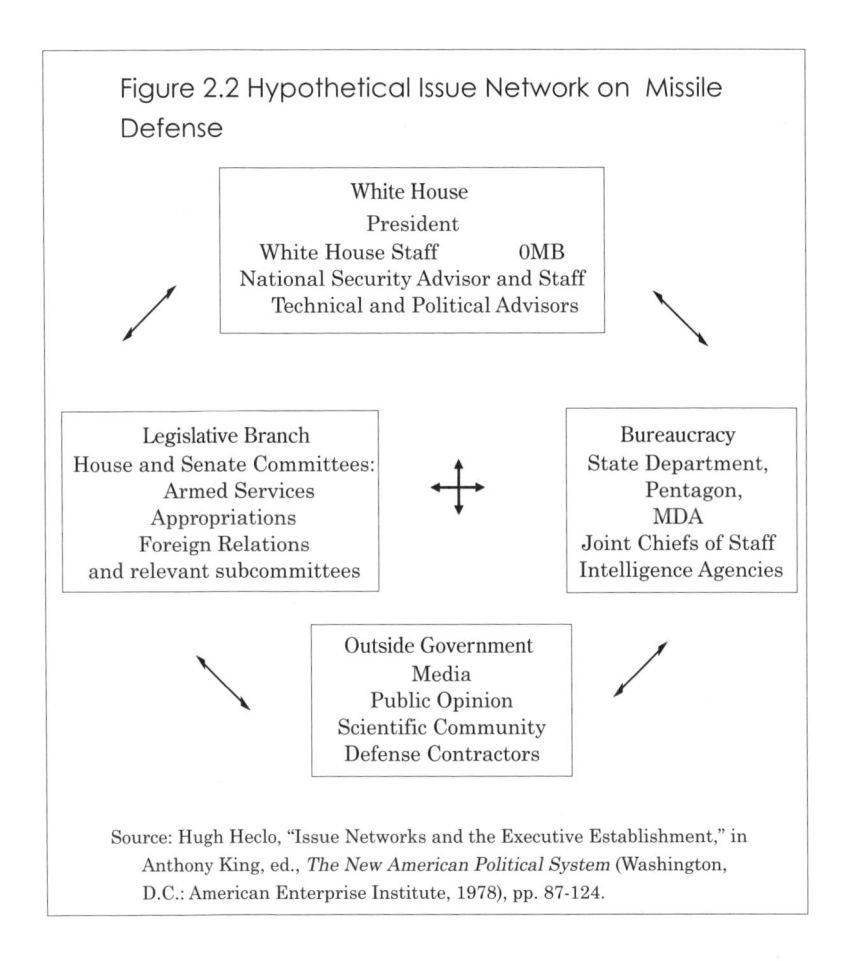

Chapter II Shaping the Missile Defense Agenda:1946-2005   49

> Figure 2.2 Hypothetical Issue Network on Missile Defense
>
> White House
> President
> White House Staff   OMB
> National Security Advisor and Staff
> Technical and Political Advisors
>
> Legislative Branch
> House and Senate Committees:
> Armed Services
> Appropriations
> Foreign Relations
> and relevant subcommittees
>
> Bureaucracy
> State Department,
> Pentagon,
> MDA
> Joint Chiefs of Staff
> Intelligence Agencies
>
> Outside Government
> Media
> Public Opinion
> Scientific Community
> Defense Contractors
>
> Source: Hugh Heclo, "Issue Networks and the Executive Establishment," in Anthony King, ed., *The New American Political System* (Washington, D.C.: American Enterprise Institute, 1978), pp. 87-124.

and the various missile defense contractors. It might also include journalists, commentators and the media, that cover the missile defense issue; research bureaus and "think tanks," which gather and analyze data on missile defense; other committees of Congress, such as the Intelligence Committees and Foreign Relations Committees, which have an interest in the missile defense issue; other branches of the bureaucracy, such as the State Department, the National Security Council, and the intelligence gathering agencies; as well as a number of entities within the Executive Office of the President, such as the White House Staff, and the Office of Management and Budget, various technical and political advisors to the president and the president himself. When any significant number of these actors are involved, missile defense policymaking is not a closed system, and therefore, an adequate explanation of its workings might need to take into account the entire issue network.

C. Policy Communities

Missile defense policy does not have the limited participation, resistance to external influences, or the preoccupation with material interests that mark the policymaking areas best explained by the iron triangle model. Nor, however, does it feature the amorphousness, the wide and changing types of participation, or the lack of clarity about who is in control that characterizes the issue network. The truth, in fact, may lie somewhere in between these two poles in a different type of subsystem called a "policy community." As described by James B. Anderson, a policy community is "broader and more open in participation than an

> Figure 2.3 Continuum of Theoretical Approaches to Missile Defense Policy Analysis
>
> | Iron Triangles | Policy Communities | Issue Networks |
> |:---:|:---:|:---:|
> | x | x | x |
> | Triadic | Multifaceted | Highly Multifaceted |
> | Closed | Moderately Open | Highly Open |
> | Clear Leadership | Shifting Leadership | No Clear Leadership |
> | Actors motivated by material rewards | Actors motivated by mixed material and policy goals | Actors motivated by policy goals |
>
> Source: Developed by the author.

iron triangle but less amorphous and under more identifiable control than an issue network."[9] Figure 2.3 above illustrates the relationship among these three theoretical approaches. We can identify at least four variables that differentiate the three. The number of actors in an iron triangle is three, but that number increases as we move to a policy community and then to an issue network. Participation in an iron triangle is closed, but becomes increasingly open as we move along the continuum. In an iron triangle, there is clear policy leadership, but again, as we move along the continuum, we find that policy leadership may shift in a policy community and be nearly unidentifiable in an issue network. Finally, in an iron triangle the actors are motivated by material rewards, but policy communities feature actors motivated more by policy goals, and in issue networks policy goals may be the predominant motivation of the actors.

If we apply Anderson's idea to missile defense policy, we may find that the missile defense community includes primarily the same congressional, interest group and bureaucratic actors we would expect to find in the iron triangle relationship in addition to many but not all of the governmental and non-governmental actors who would participate in an issue network. This community would exercise considerable influence over the nature and implementation of missile defense policy as long as no important change in that policy is likely to occur. Any such change would involve a change of a greater or lesser degree in the policy agenda of American government, and would introduce new actors and influences in the policy community. Accompanying this would be any of a variety of external influences including new presidential leadership, events in foreign affairs or foreign policy, new legislative priorities, or an altered view of American defense needs.

In the next section of this chapter, I will examine the history and development of missile defense policy in an effort to determine which of these models best describes policymaking in this area. Specifically, I will be looking for the degree to which missile defense policymaking is an open or closed system, the range of actors who exercise influence over the nature and implementation of missile defense policy, the degree to which material interests are served in missile defense policymaking, and the degree to which it is possible to identify who is in control of missile defense policy. In a later section of this chapter I will evaluate the three theoretical approaches to determine which best explains missile defense policy.

2. History and Development of Missile Defense
A. Post-World War II, the ABM Treaty, and Era of Détente, 1946-1983
a. Origins of Missile Defense

The missile age began in 1944 when the first German V-2 rocket was launched against the British in World War II, terrorizing London neighborhoods. During the next year the Allies developed timed anti-aircraft artillery barrages to defend against the V-2's, but never implemented the strategy because of the fear of unexploded artillery shells falling back on the city.[10] After the war a delegation of American military officers went to Europe to study the use of ballistic missiles as weapons, and recommended that the US undertake research and development of a range of anti-missile defense systems. As early as the end of 1953, Soviet planners, envisioning the future of intercontinental ballistic missiles (ICBM's), initiated an anti-ballistic missile (ABM) program.[11] In the US in 1945, researchers at Bell Telephone Laboratories used an analog computer to simulate 50,000 intercepts of ballistic missile targets and concluded that missile defense was technologically feasible.[12] Soon after this, the US Army went to work on an anti-missile project labeled Nike-Zeus, which involved the use of radars to detect enemy warheads and guide nuclear-armed missiles to destroy the incoming missiles with nuclear explosions at a very high altitude.[13]

Within the Pentagon, this concept met with opposition from those, primarily in the US Air Force, who favored an approach that emphasized shooting down the enemy missiles shortly after launch in their boost phase. Although the next decade of research, experimentation and testing produced little in the way of successful results, both the US and the Soviet Union built ABM systems, which they began deploying during the 1960's.[14] The Soviet system, called Galosh, defended the city of Moscow from US ICBM's. The US system evolved from the original controversy within the Pentagon between the Army and the Air Force. It employed two types of nuclear-tipped interceptors. One of these was called Spartan, and the other was called Sprint. The Spartan interceptor was designed to attack warheads above the atmosphere while Sprint exploded those that eluded Spartan within the atmosphere. The systems remained highly costly and of dubious value because, first of all, they required a nuclear explosion at high altitude to accomplish their purpose, and secondly, they could be defeated by simply launching a significant number of offensive missiles against them.[15]

By the mid-1960's, during the administration of President Lyndon Johnson, the missile defense policymaking process might be described as most closely resembling an iron triangle with the Pentagon, the congressional armed services committees and the corporate builders of the weapons systems. Public doubts were generally at a minimum as survey research showed that the vast majority of Americans believed in the ability of scientists to develop a missile defense given enough resources and a sufficient effort.[16]

If this were an iron triangle, however, it would be an iron triangle in transition. Presidential leadership was a necessary ingredient to the success of missile defense as a policy. The Pentagon continued to feel it necessary to lobby the Johnson Administration to support missile defense, arguing that some sort of defense against the increasing number of Soviet missiles targeting the US was important to voters. Pressure upon the president also came from both the House and Senate Armed Services Committees where

Democratic allies of the president warned that an "ABM gap" between the US and the Soviets which favored the USSR could become a potent issue in the 1968 elections. Reportedly, before President Johnson decided not to run for reelection in 1968, he worried that he would be politically vulnerable on the missile defense issue if he did not move to deploy a more extensive ABM system.[17]

This forced President Johnson into an awkward posture. In an attempt to defuse the issue, at the Glassboro Summit in 1967 between President Lyndon Johnson and Soviet Premier Alexei Kosygin, the US proposed that continued deployment of the missile defense system should end on the ground that such deployments only spurred further production and deployment of offensive missiles. Mr. Kosygin, however, refused, arguing pointedly, that "Defense is moral; offense is immoral."[18] This pushed the president into the arms of missile defense advocates in the Pentagon and the congressional armed services committees. As the Johnson Administration came to an end in 1969, missile defense using nuclear-tipped warheads to bring offensive nuclear missiles out of the sky by exploding in their vicinity was an important element of US defense policy.[19]

b. The ABM Treaty and Era of Détente

During the 1960's several influences weakened and penetrated the missile defense iron triangle. One was the scientific community, which began to produce research and commentary opposing missile defense. In one study Richard Garwin and Nobel Award winning physicist Hans Bethe detailed the vulnerabilities of missile defense systems.[20] Herbert York, who had been a director of research and engineering in the Pentagon, argued that ABM systems would stimulate greater production of offensive missiles by the Soviets.[21] The Federation of American Scientists (FAS), which had a membership of approximately twenty-five hundred scientists, provided organized opposition to missile defense deployment.[22] By 1969, scientists were appearing before committees of Congress to express these views. Another was growing public concern that missile defense sites near American cities might make them bigger targets of Soviet attack, or that in case of attack the interceptors would explode at such low altitudes that they would destroy the very cities they were intended to protect. Political opposition grew to such a level that when President Richard Nixon proposed a new and more extensive Safeguard anti-missile system in 1969, it was necessary for Vice President Spiro Agnew to cast a tie-breaking vote in the US Senate in order to pass it.[23] A range of opposing forces turned what might have been an iron triangle on missile defense in the fifties and early sixties into a fairly contentious issue network by the seventies.

The most significant impact on missile defense policy during this period, however, was diplomacy. In 1972, the Strategic Arms Limitation Talks (SALT I) between the US and the Soviet Union produced a five-year freeze on deployment of strategic launchers, and also yielded the Anti-Ballistic Missile (ABM) Treaty, which limited each country to two anti-missile installations.[24] Two years later continuing negotiations led to further reductions to one installation and one hundred interceptors in each country. Through these agreements the US made a clear choice to rest its national security on the concept of deterrence. The US rejected missile defense, except for the minimal defenses allowed under the ABM Treaty, on the ground that it would continue to feed a costly and dangerous arms race between the two cold war adversaries.

The only missile defense site the ABM Treaty permitted the U.S. was called Safeguard, located at Grand Forks, North Dakota. In February 1976, a few months after the site became operational, Congress ordered the Pentagon to close the facility.[25] This, however, did not end the Pentagon's missile defense program. Research and development continued in an effort to produce defensive missiles that, unlike Safeguard missiles, did not require nuclear warheads in order to destroy incoming warheads.

Pentagon analysts, meanwhile, began to worry that the Soviets had achieved a first-strike capability, which gave them a strategic advantage over the US. This led the Joint Chiefs of Staff to recommend to President Ronald Reagan early in 1983 that the US should place a greater investment in missile defense.[26]

B. The "Star Wars" Era, 1983-1993
a. The Strategic Defense Initiative

President Reagan, working largely in secret, drafted a portion of a national address on defense policy, which he delivered on March 23, 1983. In it he called upon the scientific community to employ its talents toward rendering nuclear weapons "impotent and obsolete," and pledged a "comprehensive and intensive" effort toward that end. His words caught many of his own advisors by surprise, including Secretary of State George Schultz, but they launched the largest and costliest military research and development program in US history.[27] An advisory panel under the direction of James Fletcher, a former chief of the National Aeronautics and Space Administration (NASA), began work on the technical concepts. Another panel under the direction of Fred Hoffman from a think tank, which was called Panheuristics, worked on political and strategic aspects of the policy.[28] Within a year after the President's speech, President Reagan was ready with a $1.8 billion budget for what was labeled the Strategic Defense Initiative (SDI), also known as "Star Wars."[29] The plan included a new agency to be called the Strategic Defense Initiative Organization (SDIO) in order to avoid tangling its work in the Pentagon bureaucracy. President Reagan's popularity helped to persuade Congress to accept his opinion on the scale of the missile threat from the Soviet Union, and Congress responded with a budget close to the President's request. With this accomplished, Pentagon planners were once again able to place missile defense into the center of US security strategy.[30]

After President Reagan's reelection victory in 1984, two factions grew up within the Reagan Administration on missile defense policy. One, led by Secretary of State Schultz, wanted to use missile defense as a bargaining chip in order to negotiate reductions in offensive weapons from the Soviet government. The other, headed by Secretary of Defense Caspar Weinberger, saw missile defense as a way of getting rid of the ABM Treaty, ending arms control accords, and gaining clear military superiority over the Soviets. President Reagan sided with neither group, and instead continued to believe that missile defense would render nuclear weapons obsolete.[31]

All missile defense advocates, however, a greed on two

things. The first was that SDIO needed to arrive at a single overall missile defense strategy, and the second was that missile defense would require steadily increasing levels of funding from Congress for future fiscal years. The single strategy, which crystallized in 1987, was based on the threat of a massive Soviet missile attack. Labeled Strategic Defense System (SDS) Phase I Architecture, it consisted of six major subsystems: a space-based interceptor, a ground-based interceptor, a ground-based sensor, two space-based sensors, and a battle management system. Two years later, the space-based interceptor was replaced by a concept known as "Brilliant Pebbles," which was developed at Lawrence Livermore Laboratory. This system consisted of several thousand small interceptors that were difficult to track orbiting the earth in a constellation that would cover appropriate regions of the world. Meanwhile, SDIO's budget was increasing at the same rate as the technological sophistication of its plans. For fiscal 1987, the SDIO budget was $5.4 billion, a 77 percent over the $3.1 billion that had been budgeted in fiscal 1986 and triple the original budget of $1.8 billion for fiscal 1985.

b. A "New World Order"

By the time the new Brilliant Pebbles strategy was accepted in theory, it had already rendered obsolete by events in the international system. By the end of 1989, the Soviet government was on its way to total collapse thus ending the threat for which all missile defenses up to that time had been devised. New threats, however, readily emerged. Ambassador Henry F. Cooper, the chief US negotiator at the Defense and Space Talks in Geneva, Switzerland, reported to President George H. W. Bush in March 1990 that in the post-Cold War world, the most serious threat to the US would come in the form of unauthorized or terrorist attacks made up of rather small numbers of missiles, not the thousands of Soviet warheads for which all previous strategies had been preparing. Furthermore, Ambassador Cooper advised that, as missile technology proliferated among nations previously lacking in advanced weapons programs, deployed US forces would increasingly face threats from short-range theater missiles.

Only a few months later, in August 1990, Iraqi forces invaded Kuwait, an attack that precipitated the Persian Gulf War of 1991. In this conflict, Iraq used short-range Scud missiles to attack targets in Israel and Saudi Arabia, including a US barracks in Dharan, Saudi Arabia, which killed 28 American soldiers and injured another one hundred. The US response to the Scuds was to rush Patriot missiles into service—missiles which had been originally designed to defend against aircraft. Initially, the Pentagon and the Raytheon Corporation, the manufacturer of the Patriot, claimed a near perfect success rate against the Iraqi Scuds. Such claims were motivated by two desires shared by both the Pentagon and Raytheon. The first was to show that missile technology had spread to regimes that represented new threats to US security. The second was to demonstrate that US missile defense technology was capable of matching the threat. The claims, however, later proved to be false. A review by the General Accounting Office in 1992 concluded that at most the Patriots had brought down only four out of forty-five Scud missiles by exploding in their vicinity, and that many of the Scuds that had apparently been destroyed by Patriots had actually broken up in midair due to their own faulty construction, not because of the concussive power of the defensive missile.[32]

At this point in the evolution of missile defense policy, the issue network had reached a new height in its activity level. The American public witnessed both the military threat of a "rogue regime" and the apparent success of the Patriot missile in defending against it through extensive television coverage of the Persian Gulf War. Congress, in pursuit of public opinion, passed the Missile Defense Act of 1991, which required the Pentagon to "aggressively pursue the development of advanced theater missile defense systems, with the objective of... deploying such systems by the mid-1990's" and called for the deployment of a treaty-compliant anti-ballistic missile defense system at a single site as the initial step toward the deployment of an anti-ballistic missile system."[33] Meanwhile, missile defense opponents developed evidence to show that missile defense technology, after more than forty years of planning and development, had not solved the necessary technological problems. The missile defense issue was now highly polarized, and debate across the issue network reached a new level of intensity.[34]

C. President Clinton and Theater Missile Defense (TMD), 1993-2001

a. The End of "Star Wars"

At the time President Bush left office, the five-year projected budget for missile defense was $39 billion. The arrival of President Bill Clinton in 1993 meant a reduction in the executive branch commitment to the program by more than fifty percent to $18 billion. This reduction was the result of a Bottom Up Review (BUR) conducted by the new President in collaboration with his Secretary of Defense and former chairman of the House Armed Services Committee, Les Aspin. Secretary Aspin was convinced that President Reagan's SDI had been useful in bringing about the end of the Cold War despite the fact that President

Reagan's plan had never gotten off the drawing board. Now that the Cold War was ended, however, Secretary Aspin saw no reason to continue funding a space-based missile defense system to defend against massive attacks upon the US from the world's only other superpower. The new dangers in the international system were to be found among emerging regional powers, such as Iraq, that might seek to acquire offensive missiles in order to dominate neighboring states and to challenge US interests in the area.

This was not the only problem for the new administration. As Armed Services Committee Chairman, Mr. Aspin had been aware that for many years during the Cold War the members of that committee had shored up the sagging economies of their districts by building new military bases in them, or by expanding and modernizing older ones. He was also aware that by the mid-1990's, many of those bases had outlived their usefulness and had become too costly to maintain. Some of them would have to be closed, but the decision to close them would be a bitter pill to swallow. Lucrative contracts for new work on high technology weapons, such as ballistic missile defense, however, could be one way of managing the transition and sweetening the pill. While many Democrats criticized SDI for its high cost and dubious feasibility, Mr. Aspin may have realized that money spent on research and development on missile defense could produce private sector employment for many people in at least a few of the places hard hit by base closings.

Taking these factors into account, Secretary of Defense Aspin assigned top priority to missile defense while also declaring an end to the Star Wars concept. In its place he instituted a new missile defense program, which would feature the ability to rapidly deploy shorter range missiles to regional conflict zones and defend US allies against threatening neighbors. Secretary Aspin renamed the SDIO the Ballistic Missile Defense Organization (BMDO), and projected a $12 billion budget over five years to three major projects in the area of theater missile defense (TMD).[35] The first was a new Patriot missile system, Patriot Advanced Capability-3 (PAC-3); the second was a sea-based system with missiles mounted on Aegis-class destroyers called Navy Area Defense (NAD); and the third was designed to defend against longer range missiles and was called Theater High Altitude Area Defense (THAAD). Within a few years, two more projects acquired program status: a sea-based version of THAAD called the Navy Theater Wide program (NTW), and a land-based system intended for deployment in Europe, the Medium Extended Area Defense System (MEADS).[36]

For several years during this period, the Clinton Administration tried to persuade the Japanese government to collaborate with the US on the development of TMD projects for the regional defense of East Asia. By 1997, despite a lack of broad-based support in the Japanese electorate, the Japanese Defense Agency had concluded the NTW system, described above, would be the TMD system which would be most amenable to bilateral cooperation and the one capable of defending Japan most effectively.[37] The formal decision to go forward on the US-Japan collaborative effort to develop NTW, however, had to be postponed because of pressure from the Chinese government and a lack of consensus about the desirability of TMD in Japan. Certain figures within the Japanese Liberal Democratic party (LDP) questioned the technical feasibility and cost of TMD in general, and the NTW program in particular.[38] At the same time that this debate was going on in Japan, the Chinese were not only registering loud opposition to TMD in East Asia, but also stepping up their own rather aggressive missile program. The combination of these events raised the specter of an East Asian arms race, and political support to go forward with the US-Japan collaboration began to sag.[39]

In August, 1998, however, the North Koreans launched a Taepo-Dong 1 missile, which crossed Japanese airspace before falling harmlessly into the Pacific Ocean. Although the missile launch itself apparently was no more than a failed effort to put a satellite into orbit, the event profoundly altered Japanese perceptions of the security environment. One year after the North Korean missile launch, Japan signed a Memorandum of Understanding (MOU) with the US in Washington, D.C. establishing a collaborative relationship to build TMD systems. The MOU specified the various technologies to be developed and included approximately a $200 million Japanese contribution to research costs over the next five to six years.[40] Although the MOU seemed to bring the two countries closer together on security issues, there remained considerable misgivings in Japan over TMD deployment.

Nevertheless, the Clinton Administration, through the programs described above, continued to place a higher priority on theater missile defense than on national missile defense. In the congressional elections of 1994, however, the Republican Party gained fifty-five seats in the House of Representatives and established a party majority for themselves in that body for the first time in forty years. As a result of this unexpected reversal, President Clinton's missile defense agenda faced new challenges. The Republican Congress placed a higher priority on missile defense in general and on NMD in particular than President Clinton did. In 1998, for example, Congress added $1 billion to the

Pentagon budget request for missile defense, mostly for NMD, as part of an omnibus appropriations bill passed just before Congress adjourned that year. For much of this time, TMD funding seemed to increase as an indirect result of the debate between Congress and the Administration over NMD. Whenever Congress moved to increase NMD, the Administration responded with an increase for TMD. The overall emphasis of the Administration defense policy at this time was to hold down defense spending in order to reduce budget deficits; but in Congress, Republicans and Democrats alike sought to glean the lucrative contracts that flowed from spending on missile defense.

The political battles between Congress and the president over TMD and NMD put considerable pressure on Pentagon missile defense planners and administrative leaders. In August 1996, the BMDO acquired a new director, Air Force Lieutenant General Lester L. Lyles. General Lyles's first job was to fashion a response to congressional efforts to move NMD out of its "technology readiness" stage (which meant that it remained largely on the drawing board or in the testing stage) to a "deployment readiness" position (which meant readiness for action). He accomplished this by offering what was called "three-plus-three" program. Under this new approach, BMDO would support three more years of developmental work leading to a test of its systems by 1999. In three more years, the US would deploy NMD if the security threat warranted it; but if the threat did not warrant such deployment, BMDO would continue to refine the system and remain prepared to deploy it within three years from the date that such a threat appeared to be emerging.[41] In 1997, BMDO awarded a $1.6 billion contract to Boeing North American of Seattle, Washington to serve as Lead System Integrator (LSI) for the NMD program.[42] Through this compromise, the Clinton Administration hoped to retain its increasingly tenuous grip on the direction of missile defense policy; the opposition Republicans, however, continued to gain strength.

b. Assessing the Threat: The Rumsfeld Commission

Running counter to the drive for NMD within the Republican Congress in the mid 1990's was the belief within the intelligence community that no sufficient threat existed to warrant NMD. A report commonly referred to as the National Intelligence Estimate (NIE) was published in 1995, representing the joint efforts of the Central Intelligence Agency (CIA), the Defense Intelligence Agency (DIA), and the Bureau of Intelligence and Research (BIR) from the State Department.[43] The NIE stated that the consensus within the intelligence community was that "no country would build or otherwise acquire a ballistic missile in the next fifteen years that could threaten the contiguous 48 states and Canada."⁴⁴ This ignited Republican anger in Congress, which led to a call for further investigation by an independent commission. After a lengthy period of negotiation between the White House and congressional Republicans over the membership of such a commission, a nine-member commission chaired by former Defense Secretary Donald Rumsfeld was impaneled. The Rumsfeld Commission, as it was called, got down to work in January 1998.⁴⁵

After only six months work, on July 15, 1998, the Rumsfeld Commission was prepared to offer its conclusion that "concerted efforts by a number of overtly or potentially hostile nations to acquire ballistic missiles with biological or nuclear payloads pose a growing threat to the US, its deployed forces and its friends and allies."⁴⁶ The Commission conceded that these missiles would not match US missiles for accuracy or reliability, but asserted that the nations that developed them would be able "to inflict major destruction on the US within about five years of a decision to acquire such a capability (ten years in the case of Iraq)." ⁴⁷

At this moment, outside events once again commanded attention. Less than a week after the report was delivered to Congress, the Iranians flight-tested a medium-range

Shahab-3 missile, and six weeks later North Korea flew a Taepo-Dong 1 missile through Japanese airspace in a failed attempt to launch a satellite into space. These events, especially the North Korean launch, made the Rumsfeld Commission appear to have prophetic insight into the real missile threat to the US.

Congress immediately went to work on new legislation, and in July 1999, it passed the National Missile Defense Act, which made it the policy of the US to deploy NMD as soon as technologically possible. At this time, the missile defense budget for fiscal year 2000, containing a 42.8 percent cut from $1.7 billion to $965.2 million in the appropriation for NMD, had already been approved. In that same budget, the appropriation for the various TMD programs had increased from $1.5 billion to $1 .875 billion, and increase of 23.5 percent. But Congress quickly went to work on a fiscal 2001 budget, which contained a near doubling of NMD appropriations to $1.9 billion and an 8.5 percent reduction in the TMD programs to $1.7 billion. The overall budget for both NMD and TMD gained 27.9 percent in fiscal 2001 from $2.84 billion to $3.63 billion.

This chain of events shows high level of activity in the missile defense issue network from the arrival of the Republican majority in Congress in 1995 until the 2000 presidential election year. The NIE of 1995 might well have put

NMD permanently on the back burner as public policy while TMD received most of the attention. The near simultaneous arrival of a new party majority in Congress, however, led to the appointment of the Rumsfeld Commission, which quickly sought to discredit the NIE and place NMD back in a position of prominence. The Iranian and North Korean missile launches were fortuitous events from the point of view of NMD advocates. Despite the fact that these missile launches illustrated the low level of development attained by these two countries in their missile programs, NMD advocates could exploit them to achieve policy change. Although the Clinton Administration favored research and development in TMD, Congress, working through the Armed Services Committees, increased Clinton Administration budget requests for NMD.

c. Testing NMD

At the same time that Congress was assessing the threat of missile attack, the Pentagon was at work testing missile defense systems. During 1998, 1999 and 2000, BMDO officers along with engineers from the primary missile defense contractors used a Pacific atoll named Kwajalein Island as a base for testing the technical capabilities of the proposed NMD system. In 1998, the Pentagon selected the Boeing Corporation to be the Lead System Integrator (LSI) for the NMD program. This meant that Boeing would be the corporate entity responsible for coordinating the work of the various contractors developing and testing NMD systems. The Pentagon also chose the Raytheon Corporation to build the "kill vehicle," the missile that would collide with the offensive missile and destroy it.

In October 1999, soon after the signing of the National Missile Defense Act (NMDA), the Pentagon was ready for the first flight test of NMD. The test involved firing a target missile from Vandenberg Air Force base in California, which would then be attacked by an interceptor fired from Kwajalein. This initial test was successful as the kill vehicle scored a direct hit on the target missile. Soon, however, controversy arose over whether the test missile had actually detected the target warhead or whether it had been guided towards the warhead by a large balloon used as a test decoy. At the same time, a report produced by an outside review team headed by retired Air Force General Larry Welch sharply criticized the testing effort citing inadequate testing and lapses in management. Despite the apparently successful test in 1999, NMD was again under attack.

While these events were taking place, President Clinton was wrestling with the decision to deploy NMD, which was scheduled to take place in the summer of 2000. Another test of the system occurred in January 2000 similar to the first test, but this time the interceptor missed its target by seventy meters. Both the target missile and the interceptor continued on their respective courses until the disintegrated in the atmosphere leaving NMD advocates and engineers to explain the test failure. Another report critical of BMDO was issued. This time the Pentagon's Director of Operational Testing and Evaluation (DOT&E), Philip Coyle, pointed to a hasty and hurried testing schedule as one factor causing the failure. Design flaws and maintenance problems had not been adequately diagnosed before further flight-testing was attempted. The result was not only failure but excessive waste.

At this point, as President Clinton neared the date of his NMD deployment decision, there was significant opposition within the Pentagon to making that decision at an early date. This was not to be interpreted as opposition to the concept of NMD, but rather to any hasty decision to put it into place without adequate testing. Outside the Pentagon, however, opposition to the concept of NMD was growing. Among the observers raising doubts about the validity of the successful 1999 test was the Union of Concerned Scientists (UCS), which had argued that the test was rigged to produce successful results.[48]

With strong support for NMD and other missile defense programs within the Republican Party and growing opposition on the other side, missile defense became an important issue in the 2000 campaign. The narrow victory by the Republican candidate, George W. Bush, was probably not a mandate from the voters for higher missile defense spending. It did, however, bring a new missile defense agenda to the White House.

D. Missile Defense and President George W. Bush, 2001-2005

a. The Missile Defense Agenda

On December 17, 2002, George W. Bush directed his newly named Missile Defense Agency to put into the field a national missile defense system during 2004 and 2005. He tied the new system directly to the threat of terrorism as demonstrated in the attacks of September 11, 2001:

> "September 11, 2001 underscored that our nation faces unprecedented threats, in a world that has changed greatly since the Cold War. To better protect our country against the threats of today and tomorrow, my Administration has developed a new national security strategy, and new supporting strategies for making our homeland more secure and for combating weapons of mass destruction."[49]

President Bush not only placed the highest priority on

missile defense, but also demonstrated that he was in a great hurry to realize results. Despite the fact that the systems that existed at the time had "limited operational capability," according to the Director of the Missile Defense Agency, President Bush ordered that the Agency begin to deploy the systems that were available and improve them in the field. The principle thus adopted was called "evolutionary, capability-based acquisition."[50] Essentially this meant that systems would be built and tested at the same time, even if this proved to be a very expensive way of deploying the systems.

b. Funding the Agenda

In order to accomplish this aggressive new agenda, President Bush began immediately to seek higher appropriations from Congress for missile defense. One of his first official acts upon becoming president was to increase the fiscal 2002 budget request by $3.8 billion over the previous fiscal year to $8.3 billion.[51] As I will discuss in greater depth in the next chapter, this request was cut significantly to $6.3 billion by the Senate where the party majority had shifted slightly in favor of the Democrats. The House, however, with its strong Republican majority, authorized $7.9 billion. The final appropriation was $7.8 billion.[52]

The President's budget requests for the next two fiscal

years were somewhat lower, $6.7 billion and $7.7 billion respectively; but the budget for fiscal 2005 was the highest in the twenty-one years since the strategic defense initiative was announced: $9.2 billion.[53] A highly cooperative Congress has appropriated $9.1 billion.[54] Total spending on missile defense in the administration of George W. Bush now has reached $31.9 billion in four years, which compares to $26.7 billion by the Clinton Administration in eight years.[55] The spending has produced tangible results for President Bush in the form of interceptor missiles that are now being placed in underground silos at missile site in Fort Greeley, Alaska. As of the time of this writing, at least six intercetors have been lowerd into place at the Alaska cite, and two more have been deployed at Vandenberg Air Force Base, California.[56]

3. Who Controls Missile Defense Policymaking

This anecdotal history of missile defense policymaking, covering nearly sixty years, is sufficient to reveal a complex network of powerful actors and contending forces pushing and pulling either to quash missile defense as too unworkable and costly or to make missile defense a reality as an ultimate defense against US adversaries. Yet, throughout this history, missile defense has continued to move forward despite doubts about costs, technical feasibility, compliance with the ABM Treaty, the nature of the missile threat, and the impact of the deployment of missile defense on international relations. While missile defense has occasionally become the topic of fierce debate, it overcomes all setbacks and defeats and remains a viable policy. In this chapter, I am using three models of policy analysis to explain how this has happened. I've labeled these models as iron triangles, issue networks and policy communities. The basis for my comparison consists of the following factors: 1.) the degree to which missile defense policymaking is an open or closed system, 2.) the range of actors who exercise influence over the nature and implementation of missile defense policy, 3.) the degree to which material interests are served in missile defense policymaking, and 4.) the degree to which it is possible to identify who is in control of missile defense policy.

A. An Open or a Closed System?

Clearly, missile defense policy in the US was a closed system until at least the early 1960's when scholarly opinions in opposition to missile defense began to appear in scientific journals which were also accessible to the general public. Respected scientists made two accusations against missile defense. The first was that the costs of solving the technological problem of "hitting a bullet with a bullet" were excessive. The second was that missile defense increased the likelihood of an arms race in the production of offensive missiles by potential adversaries. Despite these charges, missile defense remained entrenched as a part of US military strategy at the beginning of the Nixon Administration in 1969. Although the ABM Treaty in 1972 seemed to signal the end of the missile defense effort, in reality it was only held in abeyance.

The arrival of President Reagan in 1981 signaled the revival of missile defense in the form of the Strategic Defense Initiative (SDI), which essentially made outer space a military objective and a potential base for missile defense strategy. Criticism of this policy was intense both inside and outside the government, but President Reagan managed to shelter his new strategy by creating a special agency for it within the Pentagon: the SDIO. The SDIO helped assure that, no matter how strong the opposition to missile defense, there would always be a powerful group within the Pentagon bureaucracy motivated to perpetuate a missile defense policy. As a consequence, missile defense in the form of SDI survived through the end of the Cold War despite the fact that a missile defense policy making was now much more open than it had been twenty years earlier.

President Clinton gained limited control of a missile defense policy in 1993 when his Secretary of Defense, Les Aspin, declared the end of the Star Wars era and began to redirect the missile defense agenda toward theater missile defenses (TMD) and away from national missile defense (NMD). The Clinton Administration was skeptical toward missile defense in general and wanted to reduce overall spending in this area. President Clinton found it difficult to pursue his agenda, however, as the 1994 elections produced Republican majorities in both houses of Congress. These majorities wanted to increase missile defense spending, and to build a national missile defense system. The subsequent debate over missile defense was very open, but pro-missile defense lawmakers in Congress combined with like-minded military planners in the Pentagon to give the pro-missile defense side an advantage in political resources.

B. The Range of Actors

This dimension of policy making presents a similar pattern to the one we have just discussed. In the early stages of missile defense history, the range of actors involved with the policy was quite limited, basically consisting only of the Pentagon leadership in both the army and the air force and a few corporate think tanks and manufacturing companies doing exploratory research and development. Although there were clear signs of rivalry and conflict between the branches of the military over the policy, little of it became salient in any political sense until the mid-1960's. At this point the Johnson Administration began to display doubts about the wisdom of missile defense and met opposition from congressional advocates of anti-ballistic missile defense within his own party. President Johnson, however, was unable to finesse the issue by reaching agreement with the Soviets to limit the construction of anti-ballistic missile defenses at the Glassboro Summit. At this point the range of actors involved with missile defense consisted primarily of the members of the "iron triangle" relationship: the Pentagon, the defense contractors and the senior members of the congressional armed services committees.

The successful negotiation of the ABM Treaty by President Nixon in 1972 reduced the number of actors involved in anti-ballistic missile defense to the same essential triadic group that had kept it alive during the sixties. Twelve years later, President Reagan was able to give this group a base of operations within the Pentagon in the form of SDIO, which persisted despite controversy swirling all around it. Because the armed services committees had become more diverse in their membership at this time, the range of actors involved in the making of anti-ballistic missile defense policy was much greater at this time than it had been twenty years earlier.

As the controversy evolved during the Clinton Administration to include the debate over the relative merits of TMD versus NMD, a much broader range of actors became part of the process. At this point all of the branches of the military within the Pentagon had a stake in the shaping of missile defense policy because each one had a different program that might benefit from increases in funding for either TMD or NMD. A similar fragmentation existed on the armed services committees as members of both parties divided the missile defense budget to benefit themselves. Nevertheless, there were few voices offering to oppose missile defense altogether. The TMD versus NMD alternatives framed the debate.

C. Material Interests

Material interests in an anti-ballistic missile defense policy escalated along the entire history of its development, but they became most salient during the Star Wars era, as multi-billion dollar budgets began to be projected for the completion of the entire missile defense project. Although I will explore this dimension of policy making more thoroughly in the next chapter, we can say at this point that the creation of the SDIO and its subsequent incarnations as the BMDO and the MDA helped make it possible for these budgets to grow despite concerted opposition to this spending from some sources both within Congress and outside it. During the Clinton years, the budget for missile defense was always a matter of negotiation between the administration and the members of the armed services committees in Congress. I believe that President Clinton negotiated with the BMDO regarding the initial budget proposal to Congress. We know that in most of the years that President Clinton was in office, Congress appropriated more than the President had requested. The material interests involved in missile defense increased significantly in the administration of George W. Bush as is evident from the fact that from 2001 to 2004 both budget requests and congressional appropriations doubled.

D. Who Controls the Agenda?

On only two occasions during the history of US missile defense policy, one can see a clear indication that the president has had relatively clear and unfettered control in this area. The first of these occurred when President Nixon negotiated the ABM Treaty in 1972, and the second happened when President Reagan announced SDI and the creation of the SDIO in 1984. At other times either the Pentagon or congressional leaders had a greater share of influence; but even in these instances other actors were able to bring about change. The ABM Treaty did not end ABM research and development, nor did it prevent the Reagan Administration from announcing its Star Wars initiative twelve years later. Similarly, SDI did not prevent the transformation of Star Wars into TMD during the Clinton years (although SDI's advocates would have desired its continued development even after the Cold War had ended). The Clinton Administration had a brief opportunity to secure control over a missile defense policy in 1993 and 1994, and, as I have shown, it did manage to alter the missile defense agenda during that time. As soon as the Democratic majorities in Congress evaporated in the 1994 elections, however, President Clinton's power was reduced to negotiating for more resources for TMD while having to concede resources to Republican congressional leaders for NMD.

Congressional leaders had even fewer opportunities to gain control over a missile defense policy than the presidents did throughout this period. The senior members of the armed services committees were able to convince President Johnson to avoid political vulnerability on defense issues by funding Nike-Zeus deployment in the mid-1960's. A later generation of these leaders pressured President Clinton by negotiating missile defense budgets that favored NMD and passing legislation insisting on NMD deployment. In general, however, Congress, despite its role as the keeper of the purse strings, has usually found itself in a non-controlling role on missile defense.

My analysis in this chapter suggests that the Pentagon, in conjunction with the various missile defense contractors in the private sector, has dominated the evolution of missile defense policy since World War II. Its command of technical information regarding security threats to the US and the capabilities of missile technology along with its ability to create new knowledge through research and development have helped to produce this level of control. Political leaders have usually been forced into the position of reacting to initiatives originating in the Pentagon. President Reagan institutionalized this relationship with the creation of SDIO in 1984 while subsequent presidents have only been able to change the name of the agency and alter its mission in minor ways. As a result, the missile defense bureaucracy had remained in place until the election of President George W. Bush, who has placed NMD high on his political agenda.

4. Conclusion

The three models form a continuum with the iron triangle model representing a highly closed system with a narrow range of actors exerting influence, a similarly narrow range of material interests served and relative ease in identifying those in control. The issue network occupies the opposite end of the continuum on all of these dimensions, and the policy community takes up the middle ground. My analysis reveals that missile defense policy making has been a dynamic process, and has had different configurations at different times in its development. Figure 2.4 on the next page illustrates how these configurations have changed over time during each of the four periods I have examined. Each time period that I have discussed is given separate treatment in the figure. The horizontal line for each time period represents the continuum of policy-making models with the iron triangle model on the left end, the policy community model in the middle and the issue network model on the right. Within each time period I have attempted to identify which of the three models best describes missile defense policy-making by placing a series of X's and arrows along the continuum. These allow us to indicate the changing dynamic of policy making within each time period.

The position of X(1) in Period I indicates that NMD policy making up to the decision to deploy the Nike-Zeus system during the middle of the 1960's might well be regarded as possessing most of the attributes of an iron tri-

Figure 2.4 Models of US Missile Defense Policy Making, 1944-2005

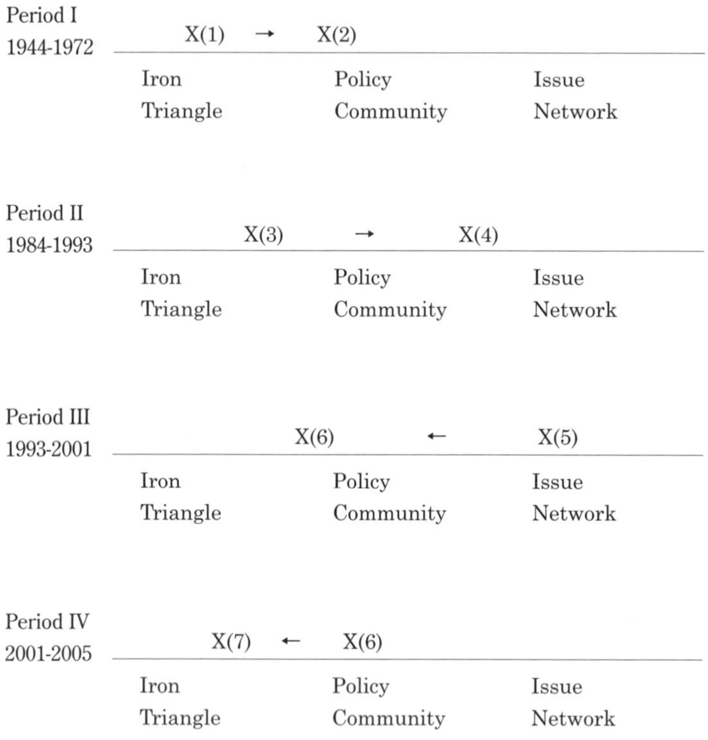

Period I
1944-1972 ──── X(1) → X(2) ────────────────────
            Iron   Policy   Issue
            Triangle   Community   Network

Period II
1984-1993 ──────── X(3) → X(4) ──────────────
            Iron   Policy   Issue
            Triangle   Community   Network

Period III
1993-2001 ──────────── X(6) ← X(5) ──────────
            Iron   Policy   Issue
            Triangle   Community   Network

Period IV
2001-2005 ──── X(7) ← X(6) ──────────────────
            Iron   Policy   Issue
            Triangle   Community   Network

Source: Created by the author

angle. Research and development of missile defense at this time was under the control of the Pentagon in conjunction with the Armed Services Committees of Congress and the defense contractors and think tanks. Although there was a dispute within the Department of Defense between the Army and the Air Force over whether enemy offensive missiles should be attacked with nuclear weapons high in the atmosphere or whether they should be targeted in their boost phase, the validity of missile defense in principle dosen't seem to have been challenged. This was a rather closed decision making process involving relatively few actors; and it was a process in which the Pentagon took the lead. Although national security was the primary motivation of missile defense advocates, the material stakes, in the form of defense budgets and highly priced military contracts, could not be denied.

The position of X(2) in that period shows that controversy over Nike-Zeus forced missile defense to move toward a policy community that included determined opponents of missile defense within the scientific community during the mid 1960's. The debate over missile defense at that time focused on the practical applications of the program and whether it could in fact improve US national security. But missile defense was also the victim of events in the international system when President Nixon and Soviet Premier Brezhnev concluded their arms control talks that led to the ABM Treaty in 1972. This limited missile defense in theory, and suppressed it in practice.

The remnants of a missile defense policy, however, fell

back into the hands of those who had kept it alive from the beginning. Although funding for missile defense from the middle to late 1970's was relatively limited, some research and development continued as part of the normal defense budget for military preparedness. During this dormant period for missile defense, the triadic relationship among congressional committees, the Pentagon and the defense contractors sustained the viability of missile defense. I judge that missile defense policy making remained static until the onset of Period II, which occurred with the announcement of SDI by President Reagan. I, therefore, place X(3) in Period II in the same location as X(2) in Period I.

The reemergence of missile defense under President Reagan as a national missile defense system was accompanied by an important political maneuver: the creation of the SDIO, which was called the BMDO under President Clinton and the MDA under President George W. Bush. Missile defense advocates in the Reagan Administration must have known that a new and more elaborate missile defense policy would immediately revitalize missile defense opponents, and that the policy making environment would be one akin to an issue network with determined and active missile defense opponents, rather than one that the Pentagon could control. SDIO insulated and protected missile defense in its incubation stage so that congressional opponents, the scientific community and public opinion could not marginalize or destroy it. The controversy over Star Wars, however, caused missile defense policy-making to more closely resemble an issue network as the debate became increasingly open. I have therefore placed X(4) in Period II along the continuum closer to the issue network side of the continuum.

Subsequent events showed that missile defense advocates had made an accurate assessment as a missile defense policy evolved through the years after the fall of the Soviet Union when the rationale for missile defense as a defense against massive Soviet missile attack disappeared and US defense doctrine had to shift completely from the bipolar world of the Cold War to fragmented and dispersed threats of the post-Cold War world. SDIO, and later BMDO, retained enough control to manage the transformation of a missile defense policy in the early 1990's. During the Persian Gulf War, despite the inadequacy of US missile defenses in that conflict, neither the Pentagon nor any congressional committee challenged the false account offered by the Raytheon Corporation that the Patriot missiles used against the Iraqi Scuds had been highly successful. Later General Accounting Office exposed the false claims.

Similarly, during the latter years of the Clinton Administration when missile defense labored under the shadow of nearly constant criticism, the BMDO persevered. With the help of a Republican majority in the House and a growing Republican presence in the Senate, missile defense retained its foothold in the missile defense budget. When the intelligence community produced a comprehensive report minimizing the missile threat from US enemies, Congress produced the Rumsfeld Commission, which derided the intelligence reports and oversold the missile threat. With the timely appearance of North Korean and Iranian missile tests, missile defense opponents had little ground to stand on as missile defense advocates boosted the missile defense budget and pushed through the National Missile Defense Act of 1999. Even while President Clinton dithered over missile defense deployment and missile tests produced little success, missile defense continued to enjoy sponsorship within Congress and the BMDO.

During Period III, therefore, I have placed X(5) in the same location that X(4) occupied in Period II to indicate the nature of missile defense policy making at the start of the Clinton Administration. But I have located X(6) closer to the iron triangle side of the continuum to reflect movement back in the direction of a more closed policy making system.

I believe that this trend has continued during Period IV from 2001 to 2005 during the administration of President George W. Bush. The President along with Secretary of Defense Rumsfeld and others in the administration have cited the war on terror as evidence of the need for a missile defense shield. They brush aside the criticism that such a shield does nothing to prevent the kinds of terrorist attacks that occurred on September 11, 2001. Instead they seem to assume that the issue is settled, and that the US will go forward with testing and deployment of missile defense systems regardless of cost or criticism.

Despite the current confidence of the Bush Administration in its missile defense policy, its future is difficult to see because so much depends on continued, general support for the Bush Administration itself. If, for example, popular endorsement of Bush's foreign policy initiatives begins to weaken, congressional willingness to fund missile defense might also begin to fall off. Nevertheless, given that the issue of missile defense has become less visible, I think it is likely that the material stakes in missile defense policy will continue to grow, and that budgets for missile defense will also increase. As the material stakes rise, the controversy over the policy will also become more intense and a larger number of actors will attempt to enter and influence the process. But even as pressure builds to widen the range of actors involved, missile defense advocates will move to protect the further evolution of a missile defense policy.

A missile defense policy is now entrenched in a relatively new bureaucracy called the MDA and, as long as it continues to enjoy the favorable attention of Republican majorities in both houses of Congress, missile defense will be the source of enormously lucrative defense contracts for high tech weapons makers. At the same time, missile defense opponents will not go away, and the issue itself will continue to have salience in an age when terrorist attacks seem a far more realistic threat to US security than enemy missiles. The future path of missile defense, therefore, will be increasingly in the hands of the president and Congress to determine. In the next chapter I will turn to an examination of certain relevant interactions between these institutions.

Notes

1 James Madison, Federalist Paper #10, in Alexander Hamilton, John Jay, and James Madison, *The Federalist Papers* (New York: New American Library, 1961), p. 79. First published in 1788.
2 Ibid., p. 83.
3 Vilfredo Pareto, *Mind and Society* (New York: Dover, 1963), p. 1476. First published in 1917.
4 Max Weber, "Bureaucracy," from *Max Weber*, trans. And ed. By H. H. Gerth and C. Wright Mills (London: Routledge and Kegan Paul, 1948), p. 226.
5 J. Leiper Freeman, *The Political Process* (New York: Random House, 1965), entire work.
6 Hugh Heclo, "Issue Networks and the Executive Establishment," in Anthony

King, ed., *The New American Political System* (Washington, D.C.: American Enterprise Institute, 1978), pp. 87-124.
7 James E. Anderson, *Public Policymaking*, 3d ed. (Boston: Houghton Mifflin Co., 1997), pp. 81-87.
8 Heclo, op. cit., p. 104.
9 Anderson, op. cit., p. 83.
10 Bradley Graham, *Hit To Kill* (New York: Public Affairs, 2001), p. 4.
11 Donald R. Baucom, T*he Origins of SDI, 1944-1983* (Lawrence, KA: University of Kansas Press, 1992), p.4.
12 Ibid., p. 2.
13 Graham, op. cit., pp. 4-5.
14 Baucom, op. cit., p. 2.
15 Ibid., p. 3.
16 Frances Fitzgerald, *Way Out there in the Blue* (New York: Simon and Schuster, 2000), p. 258.
17 Morton H. Halperin, *Bureaucratic Politics and Foreign Policy* (Washington, D.C.: Brookings Institution, 1974), pp. 297-310.
18 Graham, op. cit., p. 6.
19 Ibid., pp. 6-7.
20 Richard Garwin and Hans Bethe, "Anti-Ballistic Missile Systems," *Scientfic American*, 1968, vol. 218, no.3, pp.21-31.
21 Herbert York, "Military Technology and National Security," *Scientific American*, 1969, vol. 221(2), pp. 17-29.
22 Graham, op. cit., pp. 8-9.
23 Ibid., p. 10.
24 Treaty between the United States of America and the Union of Soviet Socialist Republics on the Limitation of Anti-Ballistic Missile Systems, Articles I-V.
25 Missile Defense Agency Historian's Office, "National Missile Defense : An Overview (1993-2000)," Missile Defense Agency,http:www.aq.osd.mil/mda/mdalink/html/mmdhist/html, p. 2.
26 Ibid.
27 Graham, op. cit., p. 14.
28 Ibid., p. 15.
29 "Historical Funding for Ballistic Missile Defense, FY85-02," Washington, D. C.: Ballistic Missile Defense Organization, 2002, p. 1.
30 Graham, op. cit., pp. 15-17.
31 Ibid., p. 15.
32 Ibid., p. 19.

33 Missile Defense Agency Historian's Office, op.cit., p.11.
34 Graham, op. cit., p. 19-20.
35 Ibid., pp. 23-24.
36 Missile Defense Agency Historian's Office,op. cit., pp. 13-14.
37 Patrick M. Cronin, Paul S. Giarra, and Michael J. Green, "The Alliance Implications of Theater Missile Defense," in *The US-Japan Alliance: Past, Present and Future* (New York: Council on Foreign Relations Press, 1999), p. 172.
38 Ibid., p.173.
39 Robert Uriu, "Japan in 1999," *Asian Survey*, vol. XL, no. 1, Winter, 2000, pp. 146-147.
40 Ibid., p.147.
41 Missile Defense Agency Historian's Office,op.cit., p.7.
42 Ibid., p.9.
43 The formal title was: "Emerging Missile Threats to North America During the Next Fifteen Years," PS/National Intelligence Estimate 95-19, November 1995.
44 Graham,op.cit., p.32.
45 The formal name of the commission was: Commission to Assess the Ballistic Missile Threat to the United States.
46 Missile Defense Agency Historian's Office,op.cit., pp.8-9.
47 Ibid., p. 9.
48 Graham, op.cit., pp. 294-316.
49 Statement by the President,"President Announces Progress in Missile Defense Capabilities," Office of the Press Secretary, December 17, 2002, http://whitehouse.gov/news/releases/2002/12/2002/217.html.
50 US Department of Defense, "Missile Defense Operations Announcement, "Missile Defense Agency News Releases,December 17, 2002,http://www.defenselink.mil/releases/2002/b12172002 bt 642-02.html.
51 Robert Snyder,"Historical Funding for Ballistic Missile Defense,Fiscal Years 1985-2005,"Ballistic Missile Defense Organization, p.1.
52 Ibid., p. 1.
53 Ibid.
54 Ibid.
55 Ibid.
56 Rick Lehner,"First Interceptor for Homeland Defense Emplaced at Vandenberg Air Force Base,California,"Missile Defense Agency : For Your Infomation, December 13, 2004, p.1.

Chapter III

The President and the Congress : Making Missile Defense Policy Choices

1. Introduction

On December 13, 2001, President Bush gave formal notice to the government of Russia that the US intended to withdraw from the Anti-Ballistic Missile (ABM) Treaty in six months. The timing of the announcement was in compliance with treaty provisions, which permitted withdrawal given six months written notice that "extraordinary events" have "jeopardized the supreme provisions of the treaty interests" of one of the parties to the treaty. The President cited no such events or interests in his announcement. He had, however, long advocated scrapping the treaty so that the US could build a land-based national missile defense (NMD) system to protect the US against future dangers from long-range missiles launched from "rogue" nations, such as North Korea, Iraq, Iran, and others.

This announcement signaled a new stage in the evolution of US missile defense policy in the post-Cold War era. President Clinton had planned to rely more heavily on several shorter range theater missile defense systems (TMD), which could be deployed quickly to defend US allies in regional conflicts. President Bush decided to move aggressively on NMD while simultaneously allowing nearly all of the TMD programs to whither away. President Bush knew that such a change in policy would require determined presidential leadership in order to gain cooperation and support in US Congress for the substantial budget allocations necessary to build the more elaborate and expensive NMD system. The shifting balance of power in Congress between the major parties during the long history of missile defense budgeting, however, has made congressional cooperation on any such plan highly problematical. Furthermore, this policy change implicated the security concerns of US allies in various danger zones around the world, such as East Asia, for example.

This chapter is about the struggle between the President and the Congress over US missile defense policy and the role of East Asian security concerns in the policy formation debate. I will use the term "missile defense" to refer to the general policy of using defensive missiles to defend against offensive missiles regardless of the specific architecture of the plan. Similarly, I will use the term "NMD" to mean the specific land-based missile defense system on US soil designed to defend the American homeland against long-range missiles. Finally, I will use the term "TMD" to indicate a category of theater missile defense systems, which are meant to be deployed in regional theaters of conflict to defend against medium and short-range missiles. As I will show in later sections of this chapter, one of the principal choices in US missile defense policy over the past fifteen years or so has been the strategic choice of whether to emphasize NMD or TMD systems in the interest of national security and the security of US allies.

The purpose of this chapter is to examine a series of questions regarding institutional interactions within the US government over missile defense policy. The first question seeks insight into the nature of presidential leadership by comparing the relative ability of President Bill Clinton and George W. Bush to secure the cooperation of Congress for their respective missile defense priorities. Essentially, I ask which of these two presidents has had greater success in gaining congressional support for his agenda. In this section of the chapter, I will include data on presidential budget requests from fiscal years 1985 through and the corresponding congressional budget allocations for those years in order to provide a context in which to assess the performance of the Clinton and Bush II Administrations. I will then use similar data from fiscal 1993 through 2005 to judge the relative success of these two administrations. The result is a comparative case study in presidential leadership.

The second question I will address amounts to this: what factors might lead a member of Congress to support or oppose specific changes in missile defense policy? I assume at the outset that the most important general factor motivating congressional behavior is the desire of the members to be reelected. Here, however, I want to know how this desire for reelection manifests itself in the area of missile defense policy. Specifically, I am concerned to know, on the one hand, whether the interests of US allies in East Asia and the security concerns of such nations as Japan, South Korea and Taiwan are significant factors in the debate over this policy. On the other hand, I want to assess the degree to which factors related to domestic political party competition and other local interests play a role in these policy choices.

In order to address this question I will look at a series of events that have put missile defense on the policy agenda of Congress and show each of them fits into a general theory of congressional reelection strategy. These events are the Rumsfeld Commission Report, the National Missile Defense Act of 1999, the national elections of 2000 and 2002, and finally, the Theater Missile Defense Improvement Act of 1998. I will reinforce the results of this historical analysis with data regarding the political links between members of Congress and defense contractors engaged in missile defense-related work for the Pentagon. Although the resulting evidence is indirect, I think it opens a useful line of inquiry into an area where direct evidence is generally unavailable.

2. Theory: Presidential Power and Congressional Motivations

A. The President

In Chapter II we found that presidential leadership and congressional motivations were key variables in the shaping of missile defense policy, especially since President Reagan's 1983 announcement of the Strateic Defense Initiative (SDI), which provided the initial impetus to the idea of developing a national missile defense system for the US. In this chapter, I begin with the assumption that any missile defense policy at all, regardless of what it contains, must be the product of shared exertions of power by the legislative and executive branches of the government. I further assume that policy leadership in this area must begin with the president of the United States as both military commander in chief and principal maker of US foreign policy. I therefore consider missile defense policy, in the first place, to be a test of presidential leadership in gaining cooperation from Congress to secure the president's preferred policy alternatives. In the second place, however, missile defense policy is also a test of congressional motivation to comply with the president's desires. I ask whether these motivations at any given time are consistent with congressional compliance with the president's preferences concerning NMD.

My analysis of presidential power in the area of missile defense begins with the recognition of the fact that the American presidency is not as powerful as many observers, including American voters, often expect it to be. These observers often fail to understand the context in which presidents operate, especially with regard to the president's chief rival power base: Congress. I take a position in general agreement with Richard E. Neustadt, which is that the American presidency is wrapped in a paradox.[1] This paradox amounts to this: although the American government centers on the presidents and the president is the one who must lead if leading is to be done, the president can rarely *command* those who follow him. Instead, the president must spend most of his time *persuading* people to join, help and vote with him. He must organize and orchestrate the motivations of others so that they will follow him because they perceive it to be in their interests to do so, not simply because the president desires it.

This view of the presidency is especially apt where Congress is concerned. Of course, the president can expect his commands to be obeyed where he is exercising his constitutional power as commander in chief of the military, where disobedience to such a command would be insubordination. Similarly, in the area of foreign affairs the president can expect a relatively free hand in setting priorities because of his dual constitutional roles as head of state and chief executive — although few presidents have had bipartisan support for their foreign policy initiatives. In the case of Congress, however, there are at least four clear sources of conflict with the president. First, members of Congress and the president represent different constituencies, the president's being national while the members of Congress have district (in the House) or statewide (in the Senate) constituencies. Second, the executive and legislative branches have different internal structures, the executive having a hierarchical and centralized structure while Congress is highly decentralized. Third, the amount and quality of information each branch may rely upon in making decisions differ markedly between the two branches, the advantage being strongly in favor of the executive. And finally, the two branches view policy from different time perspectives because of their varying terms in office. The President has a four year term and can succeed himself only once. House members have two-year terms, the Senators six-year terms; neither suffers under term limits. The average member of Congress can expect to be in office for about twelve years; but since the 1960's, American presidents have not usually been able to complete two four-year terms. Of the eight presidents elected since 1960, only three have been elected to the office twice, and only two (Reagan and Clinton) have completed two terms. Presidents, therefore, are quite sensitive to the need to make substantial policy gains early because they know that soon time will pass them by. Members of Congress, on the other hand, can usually afford to build seniority over a number of years before taking leadership over any specific policy area.[2] Therefore, where Congress is concerned, we may safely conclude with Neustadt that the power of the president is "the power to persuade."[3]

Building on Neustadt's findings, which are based on observation of specific events occurring in the Truman and

Eisenhower administrations, I assume that in order for a president to get specific action from Congress on an issue that he feels is important, at least two sets of things must happen. On the one hand, the President must state his desires clearly to Congress and make those desires known publicly so that there is no mistake in the minds of either the members of Congress or the public concerning what the president wants Congress to do. On the other hand, Congress, as the recipient of the presidential request, must be able to take the action the president desires with the view that it is operating in coordination with a necessary and proper exercise of governmental power.[4]

B. Congress

With regard to Congress, I incorporate the findings of David R. Mayhew that members of Congress are "single-minded seekers of reelection" whose actions and decisions as lawmakers are driven by their desire to increase their power and prestige in Congress so as to strengthen their reelection chances.[5] In this regard, members of Congress engage in three essential activities. The first is *advertising*, defined as getting their names before the voters they represent in a positive way. Advertising also involves dramatizing problems and getting issues on the public agenda. The second is *position taking*, which means making public, judgmental statements on matters of interest to voters. In the context of policy making, this activity would include taking a stand on the recommended solutions to any problem on the public agenda. The third and final activity is *credit claiming*, or generating the belief among the voters that they are responsible for causing some positive action by the government, which benefits them in some way. Having placed an issue on the public agenda, the member of Congress will seek credit having done something to solve it.[6]

In applying Mayhew's approach to the congressional role in missile defense policy, I further assume that members of Congress who occupy key positions on committees and subcommittees, both House and Senate, that deal directly with defense policy and defense appropriations will be most influential in shaping missile defense policy. This allows us to focus on the relatively small number of members of Congress who make the greatest difference in the shaping of that policy.

This brings us to the question of how members of Congress carry out these activities in the making of missile defense policy. If a member of Congress wants to use missile defense as a means to advertise himself to his constituents, then he must focus the attention of the public on the need for missile defense. This is what Theodore Lowi might call "overselling the remedy."[7] The member must then take a position on missile defense that marks him off from other members of Congress, but that also allows him to build the legislative coalition necessary to pass a bill on the subject. Once the bill is passed, he may then claim credit for having solved the problem. In this effort, he may be at odds with the president, even if the president is the leader of his own party. What matters most is not whether the member pleases the president, but whether his efforts have made it more likely that he will be reelected by his constituents. Indeed, the president will tolerate opposition from within his own party if the president ultimately obtains his policy preferences and the members of his party who opposed him achieve reelection to their seats in Congress. This is true because party control of Congress is ultimately more important than a few votes on any particular bill.

In the next section of this chapter I will discuss in turn the approaches of President Clinton and President George W. Bush to the problem of communicating their missile defense policy preferences to Congress. The criterion I will use to judge the leadership performance of these two presidents on missile defense will be their relative ability to secure from Congress the levels of funding for missile defense each president desired. I will discuss the political environment in which each president formulated his budget requests, and judge their performance on a comparison of the difference between their budget requests and the amounts appropriated by Congress for missile defense. This will be followed by a general evaluation of the leadership effectiveness of President Clinton and the younger President Bush.

3. Presidential Leadership
A. The Clinton Administration

From 1983 on, the missile defense issue sharply divided the parties in Congress. During the Reagan years, Democrats consistently argued that the program would be costly at a time when taxes had been reduced while the general economy seemed to be weakening. They also argued at that time that to test missile defense systems would place the US on a collision course with the 1972 Anti-ballistic Missile (ABM) Treaty and the Strategic Arms Limitation Treaty II (SALT II) between the US and the USSR. Republicans wanted to test anti-missile systems without regard for treaty constraints.

President Bill Clinton attempted to pursue a path between the two positions. After a comprehensive review of Department of Defense plans and budgets in 1993, he and

Defense Secretary Les Aspin, announced an end to the "Star Wars" era. This meant no further research into space-based missile defenses and a limited role for national missile defense systems for the US. They proposed instead to emphasize shorter-range theater missile defense (TMD) systems, which could be deployed rapidly and would be designed to defend against the type of short-range threat against which the Patriots had failed in the Persian Gulf War. Whereas the first Bush Administration had devoted 80 percent of the budget of the Strategic Defense Initiative Organization (SDIO) to national missile defense and 20 percent to theater missile defense, the Clinton Administration proposed to reverse those proportions. The result was a reduced overall investment by the US government in missile defense.

The efforts of the Clinton Administration to alter missile defense policy met with opposition after the Republicans scored an unexpectedly sweeping victory in the 1994 congressional elections, which produced Republican Party majorities in both the House and the Senate. Although missile defense had not been a national issue in that campaign, this change in party control gave the chairmanships of all key congressional committees handling missile defense appropriations to Republican opponents of the Clinton agenda. With the Republicans in control of Congress an annual struggle between the administration and various congressional committees over the budget for missile defense ensued, the result of which in most years was that Congress appropriated more money for missile defense than the Clinton Administration had proposed.

Despite the clear party divisions on the missile defense issue and Democratic control of both houses of Congress from 1987 to 1995, Congress had continued to fund missile defense procurement, research and development. Tables 3-1 and 3-2 on the next two pages show the annual presidential budget requests to cover the missile defense programs for fiscal years 1985-2001. Table 3-1 represents the nine budget requests of the Reagan and Bush I Administrations; Table 3-2 shows the eight Clinton Administration requests.[8] (Budget requests of President George W. Bush for fiscal years 2002-2005 appear in Table 3-3). Table 3-1 and 3-2 also compare those requests to the actual amounts appropriated by each house of Congress and by Congress as a whole for missile defense in those same years. The result is a snapshot of a political tug-of-war — or really two tugs-of-war: one between the Democratic and Republican parties, and the other between Congress and the president over funding for missile defense.

During the Reagan-Bush years, as Table 3-1 shows, there was a steady increase, almost year-by-year, from the first

Table 3-1. Funding for Ballistic Missile Defense, Reagan-Bush I Administrations, Fiscal Years 1985-1993
Fiscal Years, $ in Billions

	1985	1986	1987	1988	1989	1990	1991	1992	1993	Average
President's Budget Request	1.8	3.7	4.8	5.2	4.5	4.6	4.5	5.2	5.4	4.4
House Appropriation	1.1	2.5	3.1	3.1	3.2	3.1	2.3	3.5	4.3	2.9
Senate Appropriation	1.6	3.0	3.4	3.6	3.1	4.3	3.6	4.6	3.8	3.4
Final Congressional Appropriation	1.4	2.8	3.2	3.6	3.7	4.0	2.9	4.1	3.8	3.3
Difference Between Request And Final Appropriation	0.4	0.9	1.6	1.6	0.8	0.6	1.6	1.1	1.6	1.1
% Difference	22.2	24.3	33.3	30.8	17.8	13.0	35.6	21.2	29.6	25.3

Source: Robert Snyder, "Historical Funding for Ballistic Missile Defense, Fiscal Years 1985-2005" Missile Defense Agency, 2004, p. 1.

Table 3-2. Funding for Ballistic Missile Defense, Clinton Administration, Fiscal Years 1994-2001

Fiscal Years, $ in Billions

	1994	1995	1996	1997	1998	1999	2000	2001	Average
President's Budget Request	3.8	3.2	2.9	2.8	2.6	3.6	3.3	4.5	3.3
House Appropriation	2.8	2.8	3.5	3.5	3.7	3.4	3.6	4.6	3.5
Senate Appropriation	2.8	2.8	3.4	3.7	3.6	3.4	3.9	4.8	3.6
Final Congressional Appropriation	2.8	2.8	3.4	3.7	3.8	3.5	3.6	4.8	3.6
Difference Between Request And Final Difference Appropriation	1.0	0.4	-0.5	-0.9	-1.2	0.1	-0.3	-0.3	Average Diffence 0.6
% Difference	26.3	12.5	17.2	32.1	46.2	2.8	9.1	6.7	18.2

Source: Robert Snyder, "Historical Funding for Ballistic Missile Defense, Fiscal Years 1985-2005" Missile Defense Agency, 2004, p. 1.

Table 3-3. Funding for Ballistic Missile Defense, George W. Bush Administration, Fiscal Years 2002-2005

Fiscal years, $ in Billions

	2002	2003	2004	2005	Average
President's Budget Request	8.3	6.7	7.7	9.2	8.0
House Appropriation	7.9	7.4	7.5	8.7	7.9
Senate Appropriation	6.3	6.2	8.2	9.2	7.5
Final Congressional Appropriation	7.8	7.4	7.7	9.1	8.0
Difference Between Request And Final Appropriation	0.5	-0.7	0.0	0.1	Average Difference 0.3
% Difference	6.0	10.4	0.0	1.1	4.4

Source: Robert Snyder, "Historical Funding for Ballistic Missile Defense, Fiscal Years 1985-2005" Missile Defense Agency, 2004, p.1.

$1.8 billion request for 1985 to the $5.4 billion sought by President Bush (the elder) for the 1993 fiscal year. Total budget requests for these years amounted to $39.7 billion, averaging $4.4 billion per fiscal year. This represents the broad political support that had grown up within Congress for missile defense during this period. By the end of this period, many argued that spending for missile defense had been the key variable in pushing the Soviet Union to its demise and wanted to continue developing further military capabilities in order to continue to press this advantage.

The annual budget requests during the Clinton years, as shown in Table 3-2, displayed a marked drop to $3.8 billion in 1994, $3.2 billion in 1995, and then $2.9 billion, $2.8 billion and $2.6 billion in 1996, 1997 and 1998 respectively. This makes a total of $15.3 billion in budget requests for the first five fiscal years of the President Clinton's tenure. The average for these years was less than $3.1 billion per year, or approximately 25% less per year than during the Reagan Bush years.

By contrast, the final three years of the Clinton Administration showed sharp increases in the budget requests for missile defense. This may be seen as a response to severe pressure from Republicans in Congress, who had solidified and energized their newly won majorities in both houses, especially the Senate. This helps to explain the jump to a $3.6 billion request for fiscal 1999 followed by $3.3 billion and $4.5 billion for the 2000 and 2001 fiscal years. If we add the top row of figures in Table 3-2, we see that these increases brought the sum of the amounts requested by President Clinton for missile defense during both of his terms to $26.7 billion. The annual average for the eight years of President Clinton's presidency then became $3.3 billion per year, nearly 25 percent less than the average his Republican predecessors had requested.

Another indicator of the struggle between the president and Congress over missile defense is illustrated by the differences between the budget requests of the presidents and the appropriations of Congress. Throughout the Reagan-Bush years, the Democrats controlled the House of Representatives. They also held majorities in the Senate from 1987 to 1995. As long as the Democratic Party controlled majorities in both houses, they also controlled the budget process. Thus, congressional appropriations for missile defense undercut the requests of Republican presidents by an average of about $1.1 billion per year. (See Tables 3-1 and 3-2) With a Democrat in the White House and a Republican-controlled Congress, however, the shoe was on the other foot. As Table 3-2 shows, in those years — fiscal 1996-2001, Congress appropriated only about $0.2 billion more per year than it had during the Reagan and first Bush Administrations. Still, these appropriations represented $0.5 billion per year more than President Clinton requested in those years.

The second story told in these data is about the congressional response to the president's missile defense budget. For the first nine of these budget cycles, from 1985 to 1993, the Democrats controlled Congress while a Republican sat in the White House. In those years, the difference between the amount appropriated by Congress and the president's budget request for missile defense averaged 25.3% per year. The largest of these differences occurred in 1991 following the collapse of communism in Eastern Europe when Congress sought to cut federal spending by reducing defense expenditures. In that year, the president's budget request of $4.5 billion was reduced by $3.9 billion, or 35.6%.

By contrast, the budgets for fiscal years 1994 and 1995 were drawn up by the 103rd Congress during the first two years of the Clinton Administration. Despite the fact that these were the only years of unified government during the entire period from the early 1980's to the present, Congress still managed to cut President Clinton's missile defense budget request by an average of 19.6% per year. Although, as we have seen, President Clinton's requests during this period were less than those of his predecessors,

Congress was in a mood to limit defense appropriations in peacetime in an effort to get control of rapidly increasing budget deficits, and offer to the public a peace dividend in the form of cuts in the defense budget.

Republican control of Congress in 1995 gave Republicans the upper hand in budget negotiations for the 1996 fiscal year. That year proved to be the first time that Congress appropriated more money for missile defense than the president sought: a $3.4 billion appropriation compared to a $2.9 billion request, a 14.3% difference. The gap widened to 39.3% in 1997 and to 46.2% in 1998. The latter was the largest division between the president and Congress over missile defense policy since the days of Ronald Reagan. Having lost party control of Congress, the Democratic President could no longer find the political leverage necessary to keep missile defense spending in check.

In the last three budget years of the Clinton Administration, however, these differences began to narrow. Perhaps they could not have gotten any farther apart before starting to move back together; but it is more likely that outside events played a role in this convergence. By the time the fiscal 1999 budget request was about to be prepared, North Korea had tested its medium range No Dong missile. This, coupled with earlier Chinese provocations in the Straits of Taiwan, provided the occasion for missile defense supporters in Congress to issue some of their most impassioned speeches advocating a stepped up missile defense program. Missile defense foes could offer little by way of rebuttal. President Clinton's budget requests for missile defense in these last three fiscal years(1999-2001) of his administration rose to an average of $3.8 billion per year, (3.6 plus 3.3 puls 4.5 divided by 3), a 31.0% increase over the average of the previous five years. Congress, meanwhile, appropriated close to $4.0 billion per year for missile defense during those years, (3.5 plus 3.6 puls 4.8 divided by 3), a relatively small difference of only $200 million. The threatening events in East Asia combined with the fact that the Republican Party dominated both houses of Congress made it difficult for President Clinton to formulate a clear position of missile defense.

B. The George W. Bush Administration

By contrast with President Clinton, George W. Bush made missile defense an issue in the 2000 presidential election campaign, and spoke clearly on the subject. The Republican position had been developing in a consistent direction for many years, and President Bush was fully prepared to lay it before the American people. Having persuaded voters to believe in the threat of missile attack — even if it was presented in a highly exaggerated form — he was able to set forth a program voters would accept. His plan proved to be only slightly more ambitious than President Clinton's had been in terms of the number of interceptor missiles to be deployed, but far more aggressive than President Clinton's in terms of the target date for the start of missile deployment and the annual expenditure for the missile defense program. Furthermore, President Bush was confident that a Republican Congress would provide him with the resources to build the system. President Clinton, on the other hand, was forced to change his position from time to time in response to events and political pressures. His direction was uncertain, and his message was unclear.

Once again, under a new president, a name change was in order. President Bush changed the name of the Ballistic Missile Defense Organization to the Missile Defense Agency (MDA). Aside from this, the Clinton and Bush NMD proposals differed primarily in the number of interceptors, and the deployment target date each plan envisioned. The system President Clinton had under consideration in 1999 consisted of 100 interceptor missiles based in Alaska, supported by one acquisition radar and five early warning radars. Another 100 interceptors along with a second acquisition radar linked to space-based infrared sensors were also on the drawing board, scheduled to reach full operational capability by 2008.⁹

President Bush's policy envisions a much more elaborate architecture of NMD weapons systems, which include the following:

1. A total of 250 ground based interceptor missiles, half of them to be located at a base in central Alaska, the other half to be installed at the present missile site in North Dakota, which is specifically exempt under the ABM Treaty;
2. X-Band radars at two sites in Alaska, one in Greenland, another in England, and one in South Korea;
3. Upgraded early warning radars, including an additional one in South Korea;
4. A Battle Management/Command, Control and Communication headquarters to be located Cheyenne Mountain, Colorado;
5. A Space-Based Infrared System to detect missiles in their boost phase and track them throughout their trajectory;
6. An In-Flight Interceptor Communications System to pass target data from NMD sensors to interceptor missiles.¹⁰

Furthermore, President Bush would like to have it all deployed by 2005 at the latest, three years earlier than

President Clinton's plan for a more modest system. To pay for this rapidly accelerated program, President Bush Administration added $800 million to the President Clinton's fiscal 2001 budget request for the Ballistic Missile Defense Organization (BMDO), thus increasing the request to $4.5 billion. (See Table 3-2.)

Table 3-3 shows the four budget requests for missile defense that President Bush submitted to Congress for fiscal years 2002, 2003, 2004 and 2005 along with the House and Senate appropriations for those years, and the final congressional appropriation for the years for which data are available. President Bush topped the fiscal year 2001 revised request by nearly 180% by seeking $8.3 billion for BMDO for fiscal 2002, most of it earmarked for the specific systems involved in NMD. With a war in Afghanistan at hand, the Republican-dominated House trimmed $400,000 from this for an appropriation of $7.9 billion. The Senate, however, with a narrow Democratic majority, agreed to only $6.3 billion, 24.1% lower. Nevertheless, the final appropriation was just a half billion dollars below the president's request at $7.8 billion, only six percent lower.

A similar pattern occurred in the fiscal 2003 budget, submitted prior to the 2002 elections with an open-ended military commitment in Iraq and Afghanistan already underway. President Bush submitted his lowest budget for missile defense in that year, $6.7 billion. The House sought to increase that to $7.4 billion, but the Senate would only agree to $6.2 billion. The final appropriation was $7.4 billion, about 10.4 percent higher than the president's original request.

The 2002 elections changed the political dynamic for the fiscal 2004 budget process. This time the Republicans, and their pent up demand for missile defense, held majority control of the Senate. The president requested $7.7 billion, which the House sought to reduce to $7.5 billion. The Senate sought to raise the appropriation still higher to $8.2 billion, but final congressional appropriation was $7.7 billion, the exact amount the president had requested. For fiscal 2005, President Bush was able to link the war on terror in Afghanistan with his military assault on the regime of Saddam Hussein in Iraq and argued that the two together increased the need for a more rapid deployment of missile defenses. His request for $9.2 billion to fund missile defense had full support in the US Senate, but the House was able to trim $100 million from that figure yielding a final Congressional appropriation of $9.1 billion for the 2005 fiscal year.

The average difference from one fiscal year to the next in President Bush's four budget requests for missile defense and the final congressional appropriations for those years was 4.4 percent despite the fact that the president had to face opposition in the Senate in two of those four years. The average difference across President Clinton's eight budgets was 19.6 percent, and was almost exactly the same whether the Republicans or the Democrats held congressional majorities. President Clinton's middle road proved to be a lonely one on missile defense policy. He could not bring about an adequate convergence to his position. Although missile defense was not a particularly salient issue for voters in any of the elections during this period, missile defense policy was greatly affected by the changing political climate of the time.

C. Evaluation: Clinton v. Bush

President Clinton's relationship with Congress on missile defense, as well as on virtually every other issue on which he took a stand, fell victim to the high level of partisanship which has dominated American politics for many years now. Senate Republicans, for example, supported President Clinton's position on only 22 percent of the votes on which he took a position and which divided Congress during President Clinton's eight years in office. By contrast, Senate Republicans supported Democratic Presidents John Kennedy, Lyndon Johnson and Jimmy Carter on 38 percent of similar votes during the twelve years when those presidents were in office.[11] In the 1960's and in the late 1970's, Democratic presidents could often form coalitions with liberal Democrats in Congress and moderate Republicans to win their policy preferences. For President Clinton, however, as Edwards and Wayne point out, "After the Republican victory in the 1994 midterm elections, President Clinton was reduced to a largely reactive posture as the Republicans set the agenda for Congress."[12] This was exactly his posture as he tried to negotiate missile defense budgets with Republican-dominated committees during this period.

President Bush's success in shaping missile defense policy, on the other hand, was part of a larger picture in which the president took the early initiative and followed through effectively. Significant tax cutting legislation, major education reform, and federal aid to church-based social agencies were part of a strong legislative agenda that also included substantial increases in funding for missile defense. President Bush not only enjoyed a more favorable political climate than President Clinton did, but he was also able to set priorities more effectively and move more quickly toward his goals while he had the support of his own party in both houses of Congress and enjoyed generally high job approval ratings in American public opinion polls. In the matter of missile defense, his position on the need for a national missile defense system and its feasibility may have been entirely wrong, but nevertheless he spoke clearly, publicly and persuasively on the subject. The result was that Congress rewarded him with success on his legislative agenda, including large increases for NMD.

4. Missile Defense Policy in Congress
A. Advertising: Identifying the Threat

A national missile defense system can only be justified if there is a threat to US security requiring such a system to counteract it. For several years after the end of the Persian Gulf War, nothing occurred to dramatize any such threat. But in 1995 and 1996, China's firing of missiles in the Taiwan Straits coupled with reports that China was a significant factor in the proliferation of missile and nuclear weapons technology to such countries as Iran and Pakistan presented NMD advocates an opportunity to publicize the existence of a credible threat. NMD became an important Republican platform plank in the failed attempt of Senator Bob Dole to defeat President Clinton for reelection in 1996. The key sequence of events that helped to place NMD on the public agenda also began in 1996 when Congress acted by empowering a commission, led by former Secretary of Defense Donald Rumsfeld, to investigate the ballistic missile threat to the US.

The Commission did its work in 1997 and reported its conclusions to Congress in July, 1998. Roughly summarized, those conclusions were the following:

1. "Concerted efforts by a number of overtly or potentially hostile nations to acquire ballistic missiles with biological or nuclear payloads pose a growing threat to US...
2. The threat posed by these emerging capabilities (referring specifically to Iran, Iraq and North Korea)... is evolving more rapidly than has been reported by the intelligence community.
3. The intelligence community's ability to provide timely and accurate estimates of ballistic missile threats to the US is eroding...
4. The warning times the US can expect of new, threatening ballistic missile deployments are being reduced."[13]

A dominant theme of the Rumsfeld Commission's Report was the great difficulty in achieving any degree of certainty regarding the nature and seriousness of the missile threat. With regard to Russia, for example, the report traced the missile threat to "lingering political uncertainty," and a risk of accident or loss of control, which now appears small, but "could increase sharply with little warning."[14] Similarly in the case of China, the report alluded to "a range of uncertainties" which affect that country's political future, and the fact that China is now "less constrained" than it once was by fear of the Soviet Union.[15] Again, in connection with North Korea, the report cited an inability to determine the status of the development of the *Taepo Dong-2* (*TD-2*) missile program, but asserted that, because the intelligence community previously could not assess the progress of the *No Dong* program, the US "may have very little warning prior to the deployment of *TD-2*."[16]

At almost this same moment, on August 31, 1998, North Korea launched a three-stage rocket in a failed attempt to put a satellite into orbit. The first stage fell into the Pacific Ocean as the missile continued to fly over the northern tip of the main Japanese island of Honshu. The second stage fell into Pacific Ocean 1600 kilometers from the launch site, and the third stage burned out in a trail of debris stretching across the sea for 4,000 kilometers.[17] The event caught NMD advocates and opponents alike by surprise, and seemed timed perfectly to underscore the main point the Rumsfeld Commission was trying to make: we are uncertain about the threat, and that uncertainty makes the US vulnerable to attack.

This uncertainty became the core of the Rumsfeld Commission Report. The pro-NMD argument, as the Rumsfeld Commission framed the issue, rested on a syllogism having two major premises:

1. If the US is uncertain as to the probability of missile attack, then the US is vulnerable to such an attack.
2. If the US is vulnerable to missile attack, then the US should deploy NMD.

The minor premise was this:

3. The US cannot be certain as to the probability of missile attack.

The conclusion, and the message to Congress, was inescapable:

4. Therefore, the US is vulnerable and should deploy NMD.

The combined effect of the Rumsfeld Commission Report and the North Korean rocket launch created the occasion for members of Congress to present themselves in a positive light by declaring the US vulnerable to missile attack and calling for action. Senator Jeff Sessions of Alabama, a state which is home to a National Aeronautics and Space Administration (NASA) space flight center and $450 million rocket booster plant, said in speech on the Senate floor in 1999, "Our vulnerability to incoming missiles is a significant national security concern. North Korea's unexpected missile shot late last year was a wake-up call that we cannot ignore."[18] Similarly, Senator Wayne Allard of Colorado, the state in which the NMD command and control center will be located, exaggerated the missile threat in a floor speech in 1998 in these words, "...the US cannot defend itself against a single ballistic missile attack. This leaves all fifty states, especially Alaska and Hawaii, defenseless against any country that wants to threaten the US with ballistic missiles."[19] These assertions go far beyond any estimate of US vulnerability found in the Rumsfeld Commission Report. They illustrate the degree to which the advocates of a proposed policy, in this case NMD, will exaggerate the problem for which the proposed policy is being offered as a solution in order to gain support for that policy.

Exaggerated as the comments of these senators may be, however, they suggest that a domestic agenda in support of missile defense was about to take shape in the US. In the late 1980's and early 1990's, at a time when old-line industries, like steel and automobiles, were struggling to keep up with foreign competitors, the aerospace industry was prepared, with a heavy infusion of federal funds, to tackle the new technical problems presented by missile defense. Many in Congress began to see high-tech research and development contracts from the Department of Defense as a means to revitalize lagging local economies. Defense contractors for their part were more than willing to make financial contributions to congressional candidates who supported federal funding for a range of defense-related projects, including missile defense. In this climate, Congress was prepared to take the next step. On January 20, 1999, Republican Senator Thad Cochran of Mississippi and Democratic Senator Daniel Inouye of Hawaii introduced a bill entitled the National Missile Defense Act which would make it the policy of the US "to deploy as soon as (it) is technologically possible an effective National Missile Defense system capable of defending the territory of the United States against limited ballistic missile attack..."[20] Two weeks later, Republican Congressman Curt Weldon of Pennsylvania introduced similar legislation in the House of Representatives. The Senate bill had 52 co-sponsors, a majority of the body. The House bill had 97 co-sponsors. The members were now prepared to take a position and adopt a missile defense policy.

B. Position Taking and "Triangulation": The National Missile Defense Act (NMDA)

Congress had always been sharply divided over missile defense. During the 100th Congress (1987-1988) in the Senate, for example, 98.0 percent of Democrats opposed anti-ballistic missile tests while 82.2 percent of Republicans supported them. At the same time, 90.7 percent of Democrats supported compliance with the terms of SALT II compared to 81.8 percent of Republicans who opposed such compliance.[21] Although Congress had funded missile defense research and development for years, previous attempts on the part of Republicans to get the legislative body to make a clear policy statement on the subject had failed. In 1999, the Republican majorities in both houses faced only one obstacle: the possibility that Democrats in the Senate would stage a filibuster and force the withdrawal of the bill. Ending such a filibuster would require sixty votes, probably more than they could muster.

The political environment of the missile defense issue, however, continued to change. As Bradley Graham reports:

"...nearly every month another development abroad or at home underscored the proliferation of ballistic missile technology, including medium-range tests by Pakistan and Iran and underground nuclear tests by Pakistan and India."[22]

These events tended to play into the hands of Republican advocates of NMDA, raising the possibility the NMD could become a more salient issue for the 2000 campain year.

The Clinton Administration responded by reevaluating its missile defense posture. President Clinton's announcement in January 1999 that he would seek more funding for missile defense and engage in talks with the Russians aimed at modifying the ABM Treaty signaled "a swelling political sea change."[23] His purpose in doing this was to neutralize the missile defense issue for the 2000 election by means of a political strategy known as "triangulation." As practiced by President Clinton, triangulation meant identifying the traditional liberal and conservative positions of the two major parties on any issue, and then adopting a third position somewhere between the two. In the area of missile defense, President Clinton identified the traditional liberal position as the policy performance of most Democratic office holders, which was to spend less on missile defense and leave more of the federal budget to domestic, social welfare programs. The traditional Republican position, on the other hand, supported higher budgets for NMD and more rapid calendar for NMD deployment.

President Clinton's position, set out in the January 1999 announcement, staked out the territory between those two positions. By saying that he would seek more funding for missile defense and negotiate with the Russians to change the ABM Treaty, President Clinton did not commit himself to the goals of the National Missile Defense Act. Nor did his announcement place him in the position of reducing the US commitment to defending itself against potentially irrational adversaries. Instead, it placed him vaguely in the ideological center where most American voters seemed to locate themselves.

In Congress, this political maneuver had its impact when three key Democratic senators, who might otherwise have supported a filibuster against NMDA, took positions in support of the bill. First, Senator Joseph Lieberman of Connecticut, a member of the Senate Armed Services Committee, announced that he would co-sponsor the bill offered by Senator Cochran. Second, Senator Bob Kerrey of Nebraska, ranking member of the Select Intelligence Committee, having initially indicated a willingness to help organize the filibuster against the bill, decided that he, too, would support it. Finally, Senator Mary Landrieu of Louisiana, a junior member of the Senate Armed Services Committee, offered to support the bill with an amendment attached to it.

The amendment offered by Senator Landrieu required the US to continue seeking reductions in Russian nuclear forces and make appropriations for missile defense subject to the normal budget process. This amendment did two things. First, it allowed congressional Democrats to vote for NMD without appearing to support US abandonment of the spirit of the ABM Treaty as a centerpiece of arms reduction efforts because NMD could be part of ongoing arms reduction talks. Second, it meant that funding for NMD would be under the control of the same political process on congressional committees and subcommittees that governs defense appropriations every year. This would give the Democrats, as the minority party, more leverage over funding for missile defense because it would have be negotiated among party leaders on an annual basis. The amendment helped produce wide, bipartisan support for NMD in both houses. It was the final touch that made the triangulation strategy a success.

The vote in each house on the National Missile Defense Act was a resounding victory for the proponents of NMD. On the House side, the overall vote was 317 in favor to 105 opposed (with 12 not voting and one open seat). Republicans in the House voted for the bill by a margin of 214-2, with 6 not voting. The Democrats were almost evenly divided, yet the bill still garnered a majority of those voting, 103-102 (with 6 not voting). (One Independent voted against the bill). In the Senate, the vote was 97-3 with three Democrats voting against. President Clinton signed the bill into law on July 22, 1999. Although the future of NMD in President Clinton's mind continued to rest on questions of technological feasibility, public support, budgetary constraints and international acceptance, for most members of Congress it was clear that a position in support of NMD enhanced reelection chances at a time when every competitive advantage was necessary.

C. Credit Claiming: The 2000 and 2002 Elections

Votes on missile defense were not crucial to electoral success for members of Congress in either the 2000 or the 2002 elections. As Michael Barone points out in his analysis of the 2000 elections, the most salient issues dividing American voters were those of culture, religion and lifestyle, not foreign policy or defense.[24] Furthermore, American major political party competition at the national level in these elections was virtually even both in terms of voting strength in congressional and presidential elections as well as in the party balance in both houses of Congress. In the 106th Congress from 1999 to 2001, the republicans held 222 seats in the House of Representatives to 211 Democrats (2 were Independents). They lost two seats in the house in the 2000 elections(for the 107th Congress), but gained back nine in the 2002 elections. This gave them a 23-seat majority in that body for the 108th Congress.[25]

In the 106th Congress, the Republicans held a ten-seat majority, 55-45; but the 2000 elections created a rare party standoff at 50-50 with eleven new senators taking office for the 107th Congress. Since the Vice-President presides over the Senate and may vote to break tie votes, new Vice-President Dick Cheney was able to maintain a working Republican majority under a power sharing arrangement worked out with the Democratic leadership. This delicate balance tipped when Republican Senator Jeffords of Vermont decided to resign from the Republican caucus in protest over Bush Administration lobbying tactics. This made the party balance 50 Democrats to 49 Republicans with one Independent. Democratic Senator Daeschle of South Dakota became the mew Majority Leader, and all of the committees received Democratic chairs although their memberships were evenly balanced between Democrats and Republicans. The Republicans gained two seats in the Senate in the 2002 elections, and at that point enjoyed a razor thin two-vote advantage.[26]

The popular vote in recent national elections tells a similar story. In 2002 neither major party gained a majority of the total popular vote nationwide either in the House or the Senate races. The Republicans in 2002 received 49.5 percent of the total votes cast nationwide in House races and 49.4 percent of the vote in Senate races. This is similar to the results of the 2000 presidential election in which Republican candidate George W. Bush received 47.8 percent of the popular vote to 48.4 percent for Democratic candidate Al Gore. Neither major party can claim a true national majority in support of its policies.[27]

In this highly competitive environment, everything makes a difference. Although missile defense is not an issue that is of great importance in the minds of many voters, the position a candidate takes on that issue may affect enough voters to make a difference in a key election. This may have been true in the case of Democrat Mary Landrieu, who had won her seat in the US Senate from Louisiana in 1996 by fewer than 6,000 votes out of 1.7 million votes cast, a margin of less than half of one percent. Although President Clinton carried her state in the elections of 1992 and 1996, Louisiana had supported President Reagan in 1980 and 1984, and the first President Bush in 1988. Louisiana has extensive interests in military spending, and an electorate that is supportive of military causes. In 2000, Louisiana gave its eight crucial electoral votes to George W. Bush putting the Republicans in a position to target Senator Landrieu in her reelection bid in 2002.[28]

In 1998, however, Senator Landrieu traded her seat on the Agriculture Committee for one on the Armed Services Committee, which not only afforded her the opportunity to offer the key amendment to the National Missile Defense Act that drew Democrats to support it, but also to go before the voters in her home state in 2002 and claim credit for the success. In the campaign she cited the importance of military spending to the economy of Louisiana, amounting to "several billion dollars a year." She further pointed out that after only a few months on the Armed Services Committee "she brokered a major compromise that broke a five-year partisan deadlock" on the issue of missile defense. Despite strenuous efforts of Republicans to unseat her in 2002, she won reelection with 52 percent of the popular vote.[29]

Similarly, Senator Joseph Lieberman of Connecticut may also have strengthened himself politically in an indirect way as a result of his position in support of the NMDA. Although he was in no danger of defeat in his reelection bid for the US Senate in 2000, he was able to gain the nomination for vice-president and serve as running mate to then-Vice President Al Gore on the Democratic ticket that year. In general, however, it is not easy to say how a vote on missile defense will affect election results. If the larger budget for NMD can translate into jobs and contracts for local industries in certain places, it can be an electoral advantage to the member of Congress who can claim credit for obtaining them.[30]

D. Congressional Motivations and TMD

a. NMD v. TMD

In the middle to late 1990's US missile defense policy included a series of proposed programs which could be divided generally between a national missile defense program (NMD) and several theater missile defense programs (TMD). NMD, as I pointed out earlier, refers to a land-based system of interceptor missiles designed to defend the US from long-range missile attacks launched from "rogue" states. TMD refers to a series of systems some of which could be mounted on trucks while others could be launched from ships or from aircraft. Most of the TMD systems were designed to defend against missiles with ranges of no more than 1500 kilometers, but others were intended to have the capability to shoot down enemy missiles with ranges of up to 10,000 kilometers.

The Theater Missile Defense Improvement Act of 1998 will serve as another example to illustrate congressional motivations regarding votes on missile defense. This case perhaps more clearly indicates that the impact of missile defense policy on US foreign policy carries little weight in determining the vote of a member of Congress on any missile defense issue. In August 1999, the US and Japan signed a Memorandum of Understand (MOU) which began a bilateral collaboration in the research and development of

TMD technology. Although the MOU did not commit Japan to procure or deploy any weapon system which would result from this work, it nevertheless raised questions for East Asian nations regarding its purpose. In China, for example, defense analysts suspected that work done pursuant to MOU would lead to a TMD system that would extend to Taiwan, a result they opposed most vigorously.[31] Even if TMD did not extend to Taiwan, Chinese analysts worried that the MOU indicated a strengthening of the US-Japan security relationship which would threaten Chinese national interests.[32] Despite these serious foreign policy concerns, planning on TMD went ahead unabated.

The major TMD programs under consideration at that time were divided into two categories: "lower tier", or low altitude, and "upper tier", or high altitude. In the low altitude category were the Patriot Advanced Capability-3 (PAC-3) missile system, the Navy Area-wide Defense (NAD), and the Medium Extended Air Defense System (MEADS). The high altitude category included the Theater High Altitude Area Defense (THAAD) and the Navy Theater-wide program (NTW). All of these systems with the exception of NAD were designed with hit-to-kill warheads. Both the low and the high altitude navy systems were to be mounted on ships at sea. PAC-3 and MEADS could be installed on trucks, while THAAD was to be transported by aircraft.

Both PAC-3 and MEADS had a range of 1500 kilometers while NAD was limited to between 600 and 1000 kilometers. Both high altitude systems had a range of 10,000 kilometers.[33]

From the beginning, the Clinton Administration had a strong commitment to TMD, but Congress, especially after the Republican Party gained control of both houses following the 1994 elections, tended to seek increased funding for research and development on NMD. In 1998, for example, Congress added $1 billion to the Department of Defense budget request for missile defense and earmarked the majority of it for NMD. In the debate over missile defense funding, the Clinton Administration attempted to hold down defense spending in order to reduce budget deficits while Republicans and Democrats alike in Congress sought to increase spending on missile defense in order to claim credit for improving local economic conditions.

Table 3-4 on the next page shows the funding levels for US missile defense programs for the 1999, 2000, and 2001 fiscal years. The table also shows the percent change in funding over the previous year for 2000 and 2001.[34] Congressional funding for NMD fell to $965.2 million in fiscal 2000 for nearly $1.7 billion in 1999, a 42.8 percent drop. But Congress nearly doubled its support for NMD in 2001 by allocating $1,916.4 million for that fiscal year. Among

Table 3-4. U.S. Missile Defense Program Budgets
(Fiscal years, in millions of dollars)

	1999	2000	Percent Change	2001	Percent Change
National Missile Defense:	$1,687.9	$965.2	-42.8	$1916.4	98.5
Theater Missile Defense systems:					
NAD	284.6	325.2	14.3	274.8	-15.4
NTW	366.3	375.4	2.5	382.7	1.9
PAC-3	424.6	522.9	23.2	446.5	-14.9
THAAD	431.9	603.0	39.5	549.9	-8.8
MEADS	11.7	48.6	315.4	63.2	30.0
Total TMD Budget:	$1,519.1	$1,875.1	23.5	$1,717.1	-8.5
NMD + TMD Total:	$3,207.0	$2,840.9	-11.4	$3,632.9	27.9

Source: Pat Towell, "Pentagon's Chief of Testing Reinforces Bipartisan Movement to Postpone Anti-Missile System," *Congressional Quarterly Weekly Reports*, vol. 58, no.8 (Feburuary19, 2000) p. 373. Percent figures were calculated by the author and represent the percent change from the previous fiscal year's budget.

TMD programs, congressional funding for NAD averaged about $280 million over the three-year period with significant year-to-year fluctuations — up 14.3 percent in 2000, down 15.5 percent in 2001. NTW, meanwhile, grew by small consistent increments over the period from $366.3 million in 1999 to $375.4 million in 2000, and then to $382.7 million in fiscal 2001.

At the same time, however, PAC-3 and THAAD saw large increases in fiscal 2000, 23.2 percent and 39.5percent respectively followed by smaller cutbacks in fiscal 2001, 14.9 percent and 8.8 percent respectively. These two programs, one a lower tier and the other an upper tier program, received the highest level of funding over the three-year period. The combined cost of these two programs, therefore, came close to $1 billion per year. Congress gave the MEADS program the largest proportional increases in funding, from $11.7 million in fiscal 1999 to $48.6 million in fiscal 2000 and $63.2 million in 2001. The total budget for the five TMD programs went from $1,519.1 million in fiscal 1999 to $1,875.1 million in fiscal 2000, and then to $1,717.1 million in 2001. Using the figures in Table 3-4 comparing the total budget for NMD(the top row of figures in the Table), the total TMD budget, we can document the shift in congressional priorities for missile defense that occurred slightly in that period, but the NMD budget more than doubled. This not only reflected the higher costs of NMD research and development but also illustrated the growing sensitivity within Congress to the potential salience of NMD as an issue relevant to reelection opportunities in future years.

b. The TMD Improvement Act of 1998

Late in 1997, the members of the House National Security Committee (now the Armed Services Committee) were alarmed by news that North Korea had deployed the No Dong-1 missile in significant numbers.[35] They noted that the No Dong-1 had a range of about 1,000 kilometers, sufficient to threaten nearly all of Japan and the US forces stationed in Northeast Asia. They had also learned that Iran was making surprising progress on the development of two missiles, the Shahab-3, with a range of about 1,300 kilometers, and the Shahab-4 with a 2,000-kilometer range. The Shahab-3, therefore, could reach Turkey, Saudi Arabia and Israel while the Shahab-4 could threaten much of Europe.[36]

In response, the members of the Research and Development Subcommittee drafted legislation (H.R. 2786), known as the Theater Missile Defense Improvement Act, to authorize $147 million in additional appropriations to the Department of Defense for theater missile defense development. After the bill was introduced in the House in October of 1997, it was referred to the National Security Committee, which eventually approved the legislation unanimously and sent it to the floor of the House where it passed by voice vote. At the time it was introduced, the bill was also referred to the House International Relations Committee, which might have held hearings on the foreign policy implications of TMD. The International Relations Committee, however, waived jurisdiction over the bill and supported its passage exactly as drafted by the Research and Development Subcommittee.[37]

This demonstrates an important aspect of the role of Congress in making missile defense policy. An examination of data regarding specific characteristics of the congressional districts represented by the members of the Research and Development Subcommittee during that period will illustrate the point. Table 3-5 lists the members who were on the subcommittee during the 105th and 106th Congresses and ran for reelection in the 2000 campaign. The table includes the number of military bases in each district, the total numbers of military and civilian personnel employed on those bases, the presence of significant defense contractors in those districts, and the amount of campaign contributions from defense contractors to the members during the 1999-2000 election cycle.

Taken together the twenty-one members of this subcommittee represent districts containing forty-three military bases, more than a quarter of a million military personnel, and more than 138,000 civilian employees. The missile defense program does not provide additional funding for all of these bases, but the economic impact of missile defense spending on the bases that are involved in the program can be significant. Fort Bliss, Texas is one example. Nearly all of Fort Bliss is located in the congressional district represented by Silvestre Reyes, who was a member of the Military Research and Development Subcommittee at the time Congress voted on the TMD Improvement Act. Representative Reyes attached a memo to the subcommittee report on this bill which read in part:

> Fort Bliss, located in my district, trains all of the soldiers who provide air and missile defense for our military...most of the Patriot batteries are located at Fort Bliss. As such, the increased funds for PAC-3 technologies will directly affect these soldiers.[38]

Mr. Reyes's concern for the welfare of the soldiers who live and work on the military base in his district, rather than for the utility of missile defense, may be typical of other members of Congress who also have similar bases in their districts. An even more significant factor affecting congressional support for overall TMD expansion, however, may have been the campaign contributions to members of the subcommittee by major defense contractors. Table 3-5 lists the twenty-one members of the Military Research and Development Subcommittee of the House National Security Committee during the 105th Congress (1997-1998) who ran for reelection in the 2000 elections. They are divided into groups by political party, and for each group the table provides information regarding certain characteristics of each member's district. These characteristics are the number of military bases in the district, the estimated total number of military personnel on the bases in the district, the estimated total number of civilian employees on the bases in the district, and whether major defense contractors have offices or installations in the district. The column on the far right then provides the dollar amount of campaign contributions to each subcommittee member from major defense contractors during the 1999-2000 election cycle. This time period was immediately after the vote on the TMD Improvement Act and prior to the 2000 election.

As the Table 3-5 shows, total contributions to members of the subcommittee in both parties from political action committees representing defense contractors in the 1999-2000 election cycle were equal to $542,810. The twenty-one subcommittee members who voted on the TMD Improvement Act and ran in the 2000 election, therefore, received an average of $24,134 in campaign contributions from defense contractors. The twelve Republicans, representing the majority party, were given $344,960 for an average of $28,747, while the nine Democrats collected a to-

Table 3-5 Members of the Military Research and Development Subcommittee

House National Security (Armed Services) Committee
105th Congress (1997-1998)

Member	District Characteristics				1999-2000**
	N of Bases*	N of Military Employees*	N of Civilian Employees*	Presence of Major Defense Contractors*	Contractor Campaign Contributions
Republicans:					
Weldon***	0	—	—	Yes	$80,000
Kasich	1	—	2,900	No	$1,000
Bateman	4	16,000	9,000	Yes	$20,500
Hefley	4	22,100	9,000	Yes	$15,500
McHugh	1	10,200	2,200	Yes	$7,250
Hostettler	1	100	4,000	No	0
Chambliss	1	5,000	12,000	Yes	$53,500
Hilleary	0	—	—	No	0
Scarborough	5	36,000	12,000	No	$39,000
Jones	5	54,000	12,000	Yes	$47,460
Riley	2	4,800	4,900	Yes	$54,500
Gibbons	2	7,500	2,500	No	$26,250
Rep. Total	26	155,900	70,000		$344,960
				Mean for Rep's:	$28,747
Democrats:					
Pickett	5	73,000	39,000	Yes	$11,000
Abercrombie	7	16,000	9,000	No	$14,000
Meehan	0	—	—	No	0
Kennedy	2	2,000	5,000	Yes	$40,600
Blagoyevich	0	—	—	Yes	$13,000
Reyes	1	12,000	7,000	Yes	$24,250
Allen	2	1,000	8,000	No	$24,000
Turner	0	—	—	Yes	$30,500
Sanchez	0	—	—	Yes	$40,500
Dem. Total	17	104,000	68,000		$197,850
				Mean for Dem.'s :	$21,983
Grand Total	43	259,900	138,000		$542,810

Sources: *Data on military bases and presence of defense contractors in districts are found in Philip D. Duncan and Brian Nutting (eds.) *Congressional Quarterly's Politics in America, 2000: the 106th Congress* (Washington, D.C.: Congressional Quarterly Inc. 1999) pp. 14, 207, 242, 288, 380,400,436, 502, 594, 650, 832, 971,1005, 1079, 1168, 1213, 1265, 1328,1403, and 1406.
**Data on fefense contractor campaign contributions are collected by the Federal Election Commission and made available through Project Vote Smart at http://www.vote-smart.org. Campaign contribution data are complete through the 1999-2000 election cycle.
***Denotes Subcommittee Chairman.

Table 3-6. Comparison of Defense Contractor Contributions to Subcommittee Members with Major Defense Contractors in their Districts to Those without such Contractors, 1999-2000.

Contractors	N	Total	Average
With defense contractors	13	$438,560	$33,735
Without defense contractors	8	$104,250	$13,031
Difference		$334,310	$20,704

Source: Data extrapolated from Table 3-5 on page 148.

tal of $161,850 for a $17,983 average. Being a member of the majority party on this subcommittee, therefore, meant an advantage of nearly $11,000 in contributions from major defense contractors.

Regardless of party, however, members from districts where major defense contractors operated as significant employers held an even greater advantage in campaign fundraising. In Table 3-5 I identified the subcommittee members who have such contractors operating in their districts and those who do not. In Table 3-6, I compare the average amounts of total defense contractor contributions for these two groups on the subcommittee.

The thirteen members of the subcommittee representing such districts obtained an average of $33,735 from political action committees representing defense contractors. This leaves an average of $13,031 in defense contractor contributions for the eight subcommittee members representing districts where such contractors were not significant employers. The difference is more than $20,000 per member.

5. Conclusion: The Changing Balance of Power in U.S. Politics

Although President Reagan's goal of having the capability to launch missiles from outer space was never achieved, supporters of the idea argued that SDI was a factor in bringing about the downfall of the Soviet Union five years later. Whatever the validity of that claim, SDI reinforced the strategic thinking of the Secretary of Defense Caspar Weinberger, who believed that in order to gain the upper hand over the Soviet Union in Cold War diplomacy, it was necessary to demonstrate American willingness to invest US dollars in new weapons systems and countermeasures — even if they seemed patently unfeasible at first blush. In this way, Weinberger thought, the Soviets would soon become convinced that they could never gain the advantage over the US, and the result would be their Cold War capitulation and US victory.[39]

Although the threat of missile attack was greatly diminished with the end of the Cold War, it had not entirely disappeared. The fear remained that Russian offensive missiles were insufficiently secured from accidental launch or from theft by terrorists. New "Rogue" states, such as Iran, Iraq and North Korea, emerged with military ambitions of their own and the ability to threaten US allies (Israel and Japan most obviously) with weapons of mass destruction (WMD). The Persian Gulf War of 1991 demonstrated that Iraq was able to use offensive missiles as part of a bid for military dominance in the Middle East, and that short-range missile defenses needed significant technological improvements in order to become an effective countermeasure. The experience of the first President Bush in this conflict, however, was inconclusive. NMD policy was left in the hands of future presidents.

President Clinton saw the issue of missile defense as an empirical question, one that was to be decided on the basis of multidimensional evidence. NMD advocates not only had to demonstrate the technological feasibility of such a policy, but the policy itself would also have to gain support among voters in the US and among governments overseas. Otherwise, funding for the NMD program would only be enough to keep it alive from year to year. President Bush, by contrast, felt no need to address empirical questions.

He assumed the feasibility of missile defenses from the start, and had a ready-made constituency in support of the policy. Satisfied that these two aspects of the problem were resolved in his favor, President Bush could step upon the international stage to present NMD as fait accompli. This might irritate foreign leaders, but as far as President Bush was concerned the only foreign policy blunders that mattered were those that were costly at the ballot box, and he was confident that nothing of the sort would happen.

Republicans in Congress, meanwhile, were happy to follow the lead of President Geroge W. Bush, who seemed certain of where he was going. A combination of international incidents and slow political change in their favor at home over period of years dating back to the 1994 elections gave a sense of inevitability to the narrow majorities they now hold in the political branches of the US government. There may be dangers ahead, however. Nothing can be more disconcerting than to follow a confident, clear-eyed leader who suddenly discovers that he is lost. Few things can undo public policy more rapidly than the revelation that the untested assumptions underlying it are in fact false.

This may yet be the fate of missile defense in the second term of President Bush. A recent report by the General Accounting Office, the governmental agency that oversees government spending, called for the Pentagon to do more realistic and rigorous testing of NMD systems before making the systems operational.[40] The report also points out that projected budgets for missile defense total $53 billion between 2004 and 2009. The Missile Defense Agency has disputed the report, arguing that the MDA is acting in compliance with the NMDA to deploy missile defenses as soon as it can.

The Bush Administration has terminated all TMD projects except THAAD, which is the TMD system that most closely resembles the NMD system. On August 5, 2004, MDA announced that THAAD has successfully completed a "system flight certification test."[41] On July 22, 2004, the MDA announced the emplacement of the first interceptor missile at Fort Greeley, Alaska.[42] A great deal of testing, however, needs to be done on both systems. The final verdict on NMD v. TMD may come some time during President Bush's second term.

Notes
1 Richard E. Neustadt, *Presidential Power* (New York:John Wiley and Sons,1960), p.41.
2 George C. Edwards III and Stephen J. Wayne, *Presidential Leadership* (6th ed.) (Belmont, CA: Wadsworth-Thompson Learning, 2003), pp. 332-336.
3 Neustadt, op.cit., p. 20.
4 Ibid., pp.31-36.

5 David R. Mayhew, Congress: *The Electoral Connection* (New Haven, CN: Yale University Press, 1974), p.5.
6 Ibid., pp. 49-61.
7 Theodore J. Lowi, *The End of Liberalism: The Second Republic of the United States*, 2nd ed. (New York: W. W. Norton and Co., 1979), p. 139.
8 For the US Department of Defense account of the history of missile defense funding, see: Donald R. Baucom, *National Missile Defense: An Overview* (Ballistic Missile Defense Organization: Washington, D.C., 2000), pp. 1-25.
9 John Deutch, Harold Brown, and John P. White, "National Missile Defense: Is There Another Way?" *Foreign Policy*, vol. xxi, Summer, 2000, pp. 91-100.
10 Bradley Graham, *Hit to Kill: The New Battle over Shielding America from Missile Attack* (Cambridge, MA: Perseus Books, 2001), pp. 30-51.
11 Edwards III and Wayne, op.cit., pp. 338-339.
12 Ibid., p. 362.
13 US Senate, *Report of the Commission to Assess the Ballistic Missile Threat to the United States*, July 15, 1998, p. 3.
14 Ibid., p. 5.
15 Ibid., p. 15.
16 Ibid., p. 21.
17 Graham, op. cit., pp. 52-54.
18 Senator Sessions, "*US Senate Passes National Missile Defense Act*," unpublished press release, March 17, 1999, p. 1.
19 Senator Allard, "S. 187: *The American Missile Protection Ac,*" unpublished press release, Sept. 9, 1998, p. 1.
20 US Congress, *The National Missile Defense Act of 1999*, Public Law 106-38, July 22, 1999.
21 *Congressional Quarterly Weekly Report*, November 19, 1988, p. 3338.
22 Graham, op. cit., p. 105.
23 Ibid., p. 106.
24 Michael Barone, "The 49% Nation," in *The Almanac of American Politics*, 2002, Michael Barone, Richard E. Cohen and Grant Ujifusa eds. (Washington, D.C.: National Journal, 2001), pp. 21-45.
25 Milton C. Cummings, jr. and David Wise, *Democracy Under Pressure*, 9^{th}ed. (Belmont, CA: Wadsworth/Thompson Learning, 2003), p.355.
26 Ibid.
27 Ibid.
28 Michael Barone, *The Almanac of American Politics: 2002* (Washington,D.C. :National Journal Group Inc.,2002), pp.665-667.

29 Ibid., pp.666-667.
30 Ibid., pp.324-326.
31 Robert M. Uriu, "Japan in 1999," *Asian Survey*, vol. XL, no. 1 (Winter, 2000), pp. 146-147.
32 Ibid.
33 Michael O'Hanlon, "Star Wars Strikes Back: Can Missile Defense Work This Time?" *Foreign Affairs*, vol. 78, no. 6 (November/December, 2000), pp. 68-79.
34 Pat Towell, "Pentagon's Chief of Testing Reinforces Bipartisan Movement to Postpone Anti-Missile System," *Congressional Quarterly Weekly Reports*, vol. 58, no. 8 (February 19, 2000), p. 373.
35 US Congress, House Committee of National Security, *Theater Missile Improvement Act of 1998*, 105th Congress, 2nd Session, March 26, 1998, pp. 1-13.
36 Ibid., pp. 4-7.
37 Ibid., pp. 10-11.
38 Philip D. Duncan and Brian Nutting (eds.) *Politics in America, 2000: The 106th Congress* (Washington, D.C.: Congressional Quarterly Inc. 1999), p. 1329.
39 Richard Barnett, "Reflections," *The New Yorker*, March 9, 1987, p. 8.
40 General Accounting Office, "Missile Defense:Actions Are Needed to Enhance Testing and Accountability, "GAO Highlights,GAO-04-409, April 23 ,2004,p.1.
41 Missile DefenseAgency, "THAAD System Completes Important Test," For Your Information, August 5, 2004, p.1.
42 Missile Defense Agency,"Missile Defense Agency Emplaces First Interceptor at Fort Greeley," For Your Information ,July 22, 2004, p.1.

Chapter IV

The Impact of U.S. Missile Defense Policy on East Asian Security

1. Introduction

The previous two chapters have shown that missile defense has been a central component of the US response to its position in the international system after the collapse of the Soviet Union. Indeed, to the proponents of expanding this policy, the effort to develop missile defense was a key part of the strategy that led to that regime's downfall. In the post-Cold War era, the rationale for missile defense has changed, but its advocates have continued to assert that its current role is parallel to the one it played in that earlier scenario. Essentially, US willingness to develop these defensive systems raises the ante in military gamesmanship, and puts pressure on prospective US adversaries to choose between increasing their own investment in military hardware and yielding to US economic and political influence. US policy presupposes that the latter choice is the rational one for every potential foe now in existence. In the event of a non-rational choice by these foes, however, NMD proponents argue that a national missile defense system will be adequate preparation against any possible missile attack.

The previous two chapters have also shown that the contours of US missile defense policy are tightly bound up in domestic US politics. To achieve his preferences and gain congressional compliance in this and every other policy area, any US president must employ his power to persuade. Furthermore, national competition for party majorities in the two houses of Congress is now at a very high level. In this environment, even an issue like missile defense, which has little salience with voters, can make a difference in determining which party has leverage over the full range of issues coming before the government. As a result, missile defense policy, with all of its profound implications for the international system, is strongly influenced by domestic political dynamics.

In this chapter I will employ historical analysis to explore the implications of US missile defense policy as it affects the delicate security balance in East Asia. I will focus my study on the two most critical areas of conflict in the region: the Korean Peninsula and the Taiwan Strait.

On the Korean Peninsula, North Korea has created a crisis by pursuing and illicit uranium enrichment program, expelling international inspectors, withdrawing from the Nuclear Non-Proliferation Treaty (NPT) and restarting its plutonium producing reactor at Yongbyon. Meanwhile, on the shores of the Taiwan Strait, missiles from China play a major role in the threat that Taiwan faces. In both of these areas, I will address the question of whether US missile defense policy is a stabilizing or a destabilizing influence on potential conflict.

I will then turn my attention to the two most important regional powers in East Asia, China and Japan. Since China is a key figure in both the Korean and the Taiwanese conflicts, I will also survey the perceptions of China regarding US missile defense policy. Because Japan is the key US ally in East Asia, I will discuss the potential role of Japan in response to the situation in light of its important security relationship with the US. With regard to each of these two countries, I will explore how US missile defense policy affects its relationship with the US. Finally, I will explore the appropriateness of missile defense in the war against terror.

In the next section of this chapter, I will discuss my view of the current situation in international affairs. We live in a time when war among the great powers of the world is highly unlikely, and when most conflicts across international borders occur most often between and among nations that are relatively poor and less powerful. Yet, weapons of mass destruction are spreading, and many nations outside the developed world—North Korea, Iran, Pakistan, and others—now possess nuclear, chemical and biological weapons. In order to understand the impact of new weapons systems, such as missile defense, we must first discuss how wars break out and how such conflict can be deterred.

2. Theory: The Pluralistic Security Community and the Dangers it Faces

For all of the dangers that exist in the world today, some political scientists believe that it is a safer place now than it has been for most of world history. In his inaugural address as president of the American Political Science Association in 2001, Robert Jervis spoke of the fact that for many centuries, "war and the possibility of war among great powers have been the motor of international politics."[1] He went on, however, to point out that a dominant feature of the international system today is a "pluralistic security community" made up of the US, Japan and Western Europe. The pluralistic security community may be defined as a group of independent nations whose international goals and motivations are so in harmony that war among them has become literally unthinkable, by which he meant that neither the publics nor the political elites, nor even the military establishments in these countries expect war to occur among them.[2] Professor Jervis concluded that since "(t)hreatening war, preparing for it, and trying to avoid it have permeated all aspects of politics,...a world in which war among the most developed states is unthinkable would be a new one."[3] Jervis goes on to state that "given the scale and frequency of war among the great powers in the preceding millennia, this is a change of spectacular proportions, perhaps the single most striking discontinuity that the history of international politics has anywhere provided."[4]

This unprecedented era of peace among the developed nations, however, occurs amid a range of ominous threats from beyond the borders of the security community. Indeed, the terrorist attack of September 11, 2001 that destroyed the World Trade Center in New York and Pentagon demonstrated that the US is not much more secure in the twenty-first century than it had been during the era of the Cold War when the US and the Soviet Union faced each other with massive arsenals of nuclear weapons.

In addition to the terrorist threat, regional instability creates the danger of conflict in East Asia. Under certain circumstances, it is possible that the US would engage in military action to dismantle a nuclear threat from North Korea or to protect Taiwan from an attack emanating from China. In addition, the Persian Gulf region also contains areas of ongoing conflict. Currently, the US has large bodies of forces engaged in hostile action in Afghanistan as part of its war against the Al-Qaeda terror network. It also has an army of occupation in Iraq to quell the instability that arose after it forcibly deposed the regime of Saddam Hussein. The latter operation was undertaken under the mistaken assumption that the Iraqi regime possessed weapons of mass destruction sufficient to threaten the capitals of Western Europe.

Part of the US response to these threats is to accelerate development of its own national missile defense system and to recommend similar missile defense systems for its allies. This new defense framework upon which such systems appear to be based assumes that deterrence can be achieved by combining offensive and defensive forces into an international or "alliance" missile defense (AMD), which can eliminate both theater and strategic missile threats from outside the security community.[5] In order to critically evaluate the role missile defense might play in preventing future wars, it is necessary to briefly discuss the ways in which wars occur and may be prevented.

Wars, as Huth and Russet point out, rarely occur as a result of peacetime conflicts of interest and non-militarized disputes between or among nations. Rather, they usually arise from a series of escalating threats and counter-threats, which often include expansion in the amounts and types of weaponry along with changing and evolving strategies for their use.[6] Nations and their allies attempt to deter such threats through their own military preparedness, which they communicate to potential adversaries.

We may distinguish two types of deterrence: general and immediate. General deterrence focuses on conditions which give rise to military crises between rival nations from a state of affairs in which no crisis exists. Immediate deterrence concerns those factors, which determine the outcome of a crisis once it has erupted. Thus, a policy of general deterrence may be said to have failed if a challenger nation demands change in the status quo and then either threatens or initiates military action against its rival through border reinforcements, large-scale mobilization or other action designed to indicate the potential imminence of hostilities. The outcome of the resulting crisis between the two rival nations depends upon the success or failure of various measures designed to achieve immediate deterrence.[7]

There are two other types of deterrence deserving of

mention in this context: direct deterrence and extended deterrence. Direct deterrence describes the situation in which one nation seeks to prevent another from challenging the status quo of the direct deterrence relationship between them. Extended deterrence refers to the case in which the power of one nation creates an umbrella for other nations and provides the principal means for maintaining the status quo among a number of nations in a particular region.[8] In the case of East Asia, both general and immediate deterrence policies occur in the context of extended deterrence in which the US supplies the umbrella for Japan, South Korea, Taiwan, and the Philippines. The important thing about this is that in situations of extended deterrence, such as East Asia, the defender's deterrent threat is much more likely to be challenged than in cases of direct deterrence.

To illustrate let us take the example of the 1993-94 crisis between North Korea and the US over North Korea's withdrawal from the Nuclear Nonproliferation Treaty (NPT) and its refusal to allow the inspection of its nuclear facilities by the International Atomic Energy Agency. This event is an example of a breakdown in a policy of general deterrence within the extended deterrence context of the East Asian region. In that case, the tension between the US and North Korea, had reached "crisis" dimensions. The

US could not hope to guarantee the security of the Korean Peninsula and simultaneously tolerate a nuclear-armed North Korea. In order to prevent this result, the US had to explore the option of a preemptive attack against North Korean nuclear facilities. This means that the probability of military conflict—including the possibility of a US attack against North Korean nuclear reactors—could surely have been considered greatly heightened at that time. The crisis continued until late October, 1994 when an agreement, which came to be known as the Agreed Framework, was finally reached a new status quo between the two nations.[9] Under this agreement, North Korea agreed to shut down a 25-megawatt nuclear reactor capable of producing weapons-grade plutonium and to dispose of 25-30 kilograms of plutonium already produced from spent fuel.

The agreement, however, allowed the Pyongyang government substantial delays in fulfilling past promises about inspection of its nuclear weapons program, and permitted North Korea to keep intact for ten years or more the nuclear fuel enrichment facility it had previously pledged it would not possess under a 1992 agreement with South Korea. Meanwhile, North Korea continued to possess massive military force, including chemical and biological weapons and a number of intermediate range missiles with which it could challenge the peace of the region. In light of these realities, I can suggest that both the past and current US policies of general deterrence in the region have rested on a precarious footing, and that the overall danger in the current East Asian situation is considerable.

These facts lead us to review the steps by which both general and immediate deterrence policies can break down in relations among rival regional states. My discussion of this topic is based upon the scheme designed by Huth and Russet who described the process in five stages:

1. In the first stage, a dominant nation adopts a policy of general deterrence with regard to the balance of power in a particular region respecting its allies and their prospective challengers. This includes providing military assistance to its allies in the region in an attempt to deter another nation from taking steps to alter the status quo.
2. A challenger state (e.g., North Korea) makes some threat to change the status quo, such as development of nuclear or other highly dangerous weapons system. The general deterrence regime is supposed to prevent such challenges. When such a challenge occurs, we say that the policy of general deterrence has failed.
3. The defender (e.g., the US) now may decide to strengthen a commitment to an ally or allies in the region (e.g., South Korea and Japan), or it may nego-

tiate an agreeable change in the status quo with the challenger state. It may also now put its military forces on alert. The defender will have considered at this stage whether strengthening its alliances will be provocative, and whether its allies will accept any changes in the status quo which are negotiated. While the defender nation is shoring up its policy of general deterrence through negotiation, it also assumes a policy of immediate deterrence through military readiness in an effort to prevent the crisis from deepening. This is intended to coerce a favorable settlement from the challenger.

4. The challenger then decides not to retreat or yield to coercion in negotiation, but to press ahead for its desired changes in the status quo despite the immediate deterrent threat of the defender. If at this point, negotiations also fail, then the defender and its allies face a military confrontation.

5. With the arrival of the military confrontation, immediate deterrence has now failed as well, and the defender with its allies must decide whether to engage the challenger militarily.[10]

By my interpretation of these theoretical propositions, when a nation is protected by an alliance, general deterrence will fail only if a rival nation or challenger state is sufficiently motivated to issue a challenge despite—or because of—the alliance. In any event, if we follow a model of deterrence based on rationality, a challenger will be more likely to initiate a militarized dispute if it perceives that the benefits of such a dispute will outweigh the expected costs of armed conflict. In the context of extended deterrence this calculation must include an estimate of the probability that the nation which extends its umbrella of defense will actually engage in the fighting. Thus a challenger is more likely to question the status quo if it perceives that a protector nation lacks the resolve to commit itself militarily.

Furthermore, when a challenger state embarks upon a course of action which threatens the status quo and could lead to war, it typically expects that the war, if it comes, will result in a fairly quick military victory. Because a protracted war is extremely costly and can lead to the destabilization of the challenger's own regime, it is a result that is clearly to be avoided. Therefore, the challenger's estimate of the probability and cost of victory will be based on the military capabilities each side can bring to bear in the initial months of war. The relative size of standing military forces and preexisting stockpiles of weapons will be among the crucial factors in determining whether a crisis will lead to war, rather than the relative ability of two nations to mobilize their populations or to maximize their industrial capabilities for war. As Robert C. North has put it, "critical elements in an action/reaction process include each actor's capabilities, demands, contingency assertions (explicit or implicit), and applications of leverage."[11]

Missile defense, therefore, if effective, would increase the likelihood that a challenger would be unable to accomplish its purpose of achieving swift military success, thus deterring military adventures. This would have a positive effect on the maintenance of the status quo, of course; but one should not neglect the fact that negative effects are also likely. In the first place, if a nation deploys an elaborate system of missile defense, it may be perceived as an offensive threat by some other rival nation (e.g., China). In that case, the rival nation may respond with a great buildup of offensive missiles on their side. This creates the conditions necessary for an arms race. In the second place, if such an arms race results, the likelihood of misperception and miscalculation by potential challenger nations increases. Such an arms race, if its effects are felt in a particular region of the world, may produce a shifting military balance within the region, thus destabilizing the region creating dangers for all regional states.

Missile defense, therefore, represents what Richard

Rosecrance would describe as a "disruptive input" into the international system, which, given a certain set of environmental constraints, could lead to either desirable or undesirable international outcomes.[12] This signals caution in the development of missile defense as a deterrent strategy and suggests that a successful general deterrence policy must include a willingness to negotiate arms reductions and to employ economic incentives, rather than develop and deploy new weapons systems. These theoretical patterns of events should be kept in mind while examining the events I review below.

3. Impact of U.S. Missile Defense Policy
A. The Korean Peninsula

On the Korean Peninsula, US missile defense policy has been one of several elements in an unstable military and political environment. North Korea began to develop an offensive missile capacity at around the same time that the US renewed its efforts at missile defense in its Strategic Defense Initiative (SDI), known as "Star Wars." In the early 1980's, North Korea was able to obtain a 300-kolometer range Scud missile from the Soviet Union, copy its design and produce a number of them—some of which were sold to Iran for use against Iraq during the Iran-Iraq war during the 1980's. The North Korean missile program continued to grow after that resulting in a longer range version of the Scud by the late 1980's. They also developed a larger missile called the Nodong with a range of 1,000 to 1,300 kilometers capable of delivering a nuclear warhead. The Nodong is able to target all of Japan, and if sold to such countries as Iran or Libya, could also target Israel.

In 1998, North Korea tested its first multiple stage missile, the Taepo Dong 1, whose infamous flight over Japan caused alarm bells to ring from Tokyo to Washington, D.C. and beyond. It was this flight, and the elaborate missile development program which lay behind it, that gave such great impetus to the issue of national missile defense in the US. North Korea not only demonstrated that it could launch such a missile but also convinced US military planners that it could produce enough of them to be able to sell them to nations whose interests were at odds with those of the US. This led political leaders in the US to label North Korea and all of the potential recipients of its missile technology as "rogue states." North Korea's ability to engineer missiles of this type suggested that within five years, if it chose to do so, North Korea could develop a missile with sufficient range to reach the US. The response in the US was to increase funding for research and development of a large-scale national missile defense system. These successive rounds of missile rattling between North Korea and the US put Japan in the crossfire. Once again, Japan began to agonize over its role in an increasingly militarized region. North Korea had raised the stakes another time.

In addition to the missile defense issue, another element creating instability on the Korean Peninsula was North Korea's secret efforts to develop nuclear weapons. US-North Korean relations reached another critical stage in 1993-94 over the withdrawal of North Korea from the Nuclear Nonproliferation Treaty. As we have seen, that situation resulted in a new status quo brought about through negotiation. The response of the Clinton Administration to the 1998 crisis was to support increased funding for missile defense at home while engaging in dialogue with North Korea abroad. Through negotiations, President Clinton managed to secure an agreement with North Korea to curtail missile sales and to extend the freeze on its nuclear program, which had been established in 1994, but without significant provisions for verification. Still, in the waning days of President Clinton's term in office, it appeared that North Korea was about to abandon any future ambitions for nuclear weapons. For the first time in many years, there seemed to be cause for optimism regarding North Korea's relations with the US.

This optimism reached its peak as the twenty-first century began. Early in the year 2000, North and South Korea stunned the world by taking dramatic steps to reduce tensions between their two regimes. South Korea's goodwill policy toward North Korea produced an historic summit meeting in June, 2000 between the leaders of the two countries, South Korean President Kim Dae Jung and North Korean President Kim Jong Il. Both countries seemed committed to laying a foundation for an improved peace system on the Korean Peninsula. Subsequent negotiations between the two countries, however, led to frustration and stagnation. In November, 2001, talks were held in an effort to reach agreement on the disposition of separated families residing in the two countries. Despite arduous negotiation, no agreement was reached. Yet, there appeared to be a desire both in Seoul and in Pyongyang to achieve significant movement toward reunification in the interests of both countries.

As the Clinton Administration gave way to the administration of President George W. Bush, a new source of frustration for leaders in both Korean capitals became apparent. This stemmed from the sharp discontinuity between the Clinton Administration policy and that of the Bush Administration, which seemed intent on doing everything as differently as possible from its predecessor. President Clinton had sought to negotiate an agreement that would have ended North Korea's production of medium and long-range missiles as well as the export of missile technology. In return, President Clinton had offered to make a presidential trip to Pyongyang and to contribute hundreds of millions of dollars in food aid to that country; but before the deal could be completed, Clinton was out of office.[13] From the beginning, President Bush demonstrated his disapproval of his predecessor's willingness to seek improved relations with North Korea and irritated Pyongyang by refusing to continue negotiations and by issuing public statements labeling North Korea a producer of biological weapons.[14] By the summer of 2001, the Bush Administration was demanding rigorous verification of North Korea's promised halt of its missile technology and nuclear weapons development program. The verification scheme sought by President Bush would include "challenge inspections" in which American officials would have access to a range of sites on North Korea at short notice.[15]

The US missile defense policy was a special cause of difficulty. North Korea defined the Bush Administration missile defense program as a policy that is destroying peace with a "space missile alliance strategy." It bristled at being labeled a "rogue state," and charged that such rhetoric was a way to avoid confrontation with more meaningful challenges from China and Russia. North Korea also perceived missile defense as a way of altering the Japanese role in East Asia toward a stronger military posture, citing US attempts to get Japan to revise its laws relating to its Self-Defense Forces. It warned that Japanese movement toward a more active military role would have a negative impact on its negotiations with Japan over normalization of relations. Finally, and most ominously, North Korea announced its renewed efforts to develop nuclear weapons.

As the year 2002 drew to a close, the Bush Administration quietly opened official talks with North Korea by sending Assistant Secretary of State James A. Kelly to pay a courtesy call on Kim Yong Num, the president of the Supreme People's Assembly and the second-ranking official in North Korea. Reports indicate, however, that nothing of substance was discussed between the two and that most of Kelly's meetings were with lower-ranking officials.[16] In January, 2003, President Bush characterized the situation with North Korea as a "diplomatic issue, not a military issue," but then accused Kim Jong Il as "somebody who starves his own people." In the same statement, President Bush also expressed his belief that the US was the leading worldwide donor of food supplies to North Korea.[17] In retrospect, this statement now has appearance of a threat by the US to cut off food supplies to North Korea unless the

North Korean regime backed down on going forward with its nuclear weapons program. In mid-June, 2003, North Korea's response was to tell the US to "mind its own business."[18]

Meanwhile, Japan appeared to be farther down the arduous road toward normalization of relations having held a historic summit meeting between Japanese Prime Minister Junichiro Koizumi and Kim Jong Il in September of 2002. At that meeting, the North Korean leader agreed to freeze testing long-range missiles and reiterated a pledge to permit inspections of the country's nuclear sites.[19] Although Japan-North Korea talks appeared to be producing only the most grudging progress on other issues, their bilateral relationship was in better shape by the end of 2002 than that between the US and North Korea. The latter relationship was jolted by North Korea's admission in October, 2002 that it had initiated a clandestine program to produce enriched uranium despite having pledged not to do so in the 1994 Agreed Framework. This began a new round of escalating confrontations which included North Korea's withdrawal from the Nuclear Non-Proliferation Treaty, its expulsion of International Atomic Energy Agency inspectors, and a move to restart the plutonium production program which had been frozen under the 1994 agreement.

During 2003, President Bush held separate talks with South Korean President Roh My Hyun and Japanese Prime Minister Junichiro Koizumi. In each case the talks led to a joint declaration condemning North Korea's nuclear program and threatening "further steps" or "tougher measures" if North Korea did not back down. By mid-year, however, both Asian leaders found themselves facing political difficulties stemming from their support of the American president. In South Korea, President Roh's support of President Bush brought protests from members of President Roh's own party who said he had betrayed his promise to remain independent of the US.

President Bush also received words of caution at home. In the spring of 2003, the Task Force on US-Korea Policy, co-sponsored by the Center for International Policy in Washington, D.C. and the Center for East Asian Studies at the University of Chicago issued a report containing stern warnings. The Task Force expressed unanimous agreement that North Korea's resumption of plutonium production and its success in developing weapons-grade uranium enrichment capability would have a disastrous impact not only on East Asian security but on the entire global nonproliferation regime.[20] The recommendations of the Task Force included a warning that any effort on the part of the US to prevent the spread of nuclear weapons on the Korean Peninsula would fail without US cooperation with South Korea, Japan, China, Russia and the European Union. It further argued that the US could not succeed unless the nuclear issue was addressed along with four other issues: the normalization of economic and political relations with North Korea, guaranteeing the security of a non-nuclear North Korea, promoting reconciliation of North and South Korea, and drawing North Korea into economic engagement with its neighbors.[21] Furthermore, the Task Force cautioned against the kind of confrontational and unilateral approaches that had characterized Bush Administration interactions North Korea from the start. Such approaches "would not only exacerbate the nuclear crisis but also undermine United States relations with Northeast Asia as a whole, especially with South Korea, jeopardizing the future of the US-South Korea alliance."[22] Addressing the missile issue directly, the Task Force recommended that the US resume negotiations to end the further development of North Korea's missile capabilities, and the export of its missiles, missile technology, and missile components to other states.[23] The Bush Administration, however, has continued to argue that such negotiations would be yielding to blackmail and that North Korea cannot be trusted to keep its end of any bargain.

Seeming to move in both directions at once, the Bush

Administration agreed to negotiate the nuclear issue in six-nation talks involving the US, North Korea, South Korea, Japan, China and Russia. The first round of these talks took place in February, 2004, hosted by the Beijing government. At the same, however, the US continued to take a hard line, representing its own demands as non-negotiable. When, for example, the North Koreans sought assurances that the US would not engage in a preemptive war against them as an initial sign of peaceful intent, the US refused to grant any such assurances. On the eve of the talks, the North Koreans indicated to the Chinese Vice Foreign Minister Wang Yi that it was ready to abandon its nuclear ambitions and to freeze all of its nuclear activities as a step toward total abolition of its nuclear capability. The US, however, continued to demand a complete, verifiable and irreversible dismantling of North Korea's nuclear program, and maintained its position that a nuclear freeze was not enough.[24] Four days later, the talks ended without positive results.

The six nations agreed to hold a second round of talks in June, 2004, but serious differences continued to exist among them. Once again, the venue for the talks was Beijing, and again they ended without significant progress. China's chief delegate to the talks, Vice Foreign Minister Wang Yi, commented that the US and North Korean positions were still far apart, and that, "there is still a serious lack of mutual trust."[25] According to Wang Yi, North Korea said it was willing to give up all nuclear weapons and nuclear weapons-related programs in a transparent way, but would not give up its uranium enrichment program. This appeared to be the primary sticking point which stood in the way of any positive gesture from the US.

A few days after the talks ended, the US offered major economic aid to North Korea along with certain unspecified security guarantees and the easing of some of the political and economic sanctions. These concessions, however, were contingent upon North Korean agreement to shut down and seal its nuclear weapons facilities in a period of three months. The North Koreans rejected the offer saying that the three-month period was "so unscientific and unrealistic that nobody could support it."[26]

The US, however, continued to press for some kind of agreement. A few days after the North Korean rejection of the US latest offer, US Secretary of State Colin Powell met with North Korean Foreign Minister Paek Nam Sun in Jakarta, Indonesia. Secretary Powell offered to match North Korea "deed for deed" in the short term if it agreed to dismantle its nuclear weapons and halt their development. The Foreign Minister appeared to soften his own stand against the US, but lamented that here existed no trust between the US and North Korea and reiterated that North Korea wanted rewards for giving up its nuclear weapons.

Since the 2004 US presidential election campaign ended in victory for President Bush, the North Korean strategy of hoping for the President's defeat and a new administration that would negotiate with it somewhat more flexibly has ended in disappointment.[27] North Korea's strategy since the election has seemed somewhat erratic. On February 10, 2005, Pyongyang announced that it had built nuclear weapons and had no interest in returning to the six-nation talks; but on March 5, Chinese Foreign Minister Li Zhaoxing said that North Korean leaders had expressed to him that North Korea remains committed to a nuclear-free Korean Peninsula and that they would participate in the six-nation talks.[28] Further remarks of Mr. Li published in the West on the following day suggested that the Chinese had been hard at work to re-start the talks. Mr. Li reiterated North Korea's willingness to participate in them and urged both Washington and Pyongyang to show flexibility as they addressed North Korea's "legitimate concerns."[29]

The exact impact of US missile defense policy on this sequence of events is difficult to sort out. It is one of several disruptive inputs that, along with mutually stubborn and confrontational attitudes on both sides, has turned an already dangerous environment into one that is all the more so. Nevertheless, the situation on the Korean Peninsula seemed to be headed toward becoming less dangerous before the Bush Administration injected its more aggressive missile defense policy into the equation.

B. China and Taiwan

The US position on the status of Taiwan has always been curiously ambiguous. It is often the most readily available example of diplomatic double speak. On the one hand, the official position of the US is that there is but one China, not two. By inference, then, Taiwan is not an independent state. But on the other hand, the US sells arms to Taiwan just as if it were an independent state, and is committed to the defense of Taiwan should the Chinese ever take it into their heads to reclaim what it regards as a "rebel province."

The Chinese government has recently clarified its position on the Taiwan question through the terms of its new "anti-secession" law, which expresses China's determination to use "non-peaceful means"as a last resort to prevent Taiwan from establishing formal independence. Taiwan's leaders have denounced the law as a trigger for war, and the Bush Administration has counseled China to abandon it; but Wang said the anti-secession law "is both necessary and timly" to prevent further moves toward formal independence by Taiwanese President Chen Shui-bian and his pro-indepenndence Domocratic People's Party in Taipei. In apparent response to Washington's intervention, Mr.Wang Zhaoguo, Deputy Chairman of the National People's Congress Standing Committee,quoted the legislation as saying the struggle over Taiwan is "China's internal affair " and "we will not submit to any interference by outside forces."[30]

The Chinese maintain troops and an impressive array of offensive missiles at the ready in the event that such military action would be genuinely successful.

One significant aspect of recent US policy toward the defense of Taiwan has been missile defense. During the years of the Clinton Administration, this took the form of theater missile defense (TMD). The idea was that defensive PAC-3 missiles could be mounted on Arleigh-Burke class destroyers equipped with Aegis-style air-defense radars. These interceptors would be able to target China's short-range missiles and attack them in their launch phase, destroying them long before they reached their targets. This plan, initiated in 1993, encountered numerous difficulties after $2.3 billion had been appropriated to implement it. Cost overruns and various problems integrating the missile intercept system with Aegis systems led to the cancellation of the project. US military planners, however, have not entirely given up on the idea of deploying a ship-based, theater-wide system that would have applications for Taiwan. Reportedly, the Taiwanese defense ministry remains interested in acquiring Aegis technology.[31]

At the beginning of his term in office, President Bush made every effort to appear as unlike his predecessor as possible in dealing with East Asian security problems. Nevertheless, the rationale offered by the Bush Administration for its missile defense policy in Taiwan has been remarkably close to that of the Clinton Administration. James Mulvenon characterizes both administrations as relying on a rationale sometimes known as "the freedom of action" argument.[32] Under this view China, which possesses missiles capable of reaching the US, could deter the US from intervening to defend Taiwan by threatening full-scale missile retaliation, thus undermining US defense commitments to its ally, Taiwan.[33] This argument, however, is subject to a serious reality check: the Chinese ICBM force consists of approximately twenty silo-based, liquid-fueled missiles which cannot be launched on warning or launched under attack. A US first strike could probably wipe out this force before it got off the ground even without a national missile defense system. Even if this were not the case, it could hardly be maintained that China can deter US military action in defense of Taiwan.

Nevertheless, the Bush Administration has remained committed to its strong missile defense policy, which includes NMD. This commitment produced a downturn in relations between the US and China from the start of President Bush's term in office. As Dali L. Yang points out, Bush saw China in a different light than his predecessor, President Bill Clinton, who sought a strategic partnership with China. To Bush, China was a "strategic competitor" towards whom Bush adopted a unilateralist policy on every issue from national missile defense to arms sales with Taiwan.[34] For China's leaders, the Bush Administration's determination to go ahead with missile defense deployment threatened to neutralize China's small arsenal of nuclear missiles, while US weapons sales to Taiwan served to encourage Taiwan politicians to hold out against mainland overtures for national reunification.

US weapons sales to Taiwan, however, seem to be a more salient issue than NMD at the present time in US-China relations. President Jiang Zemin suggested during his meeting with President Bush in October 2002 that China could link its deployment of short-range missiles facing Taiwan to US arms sales to the Taiwanese military.[35] The offer seemed to call the US government's bluff on the arms sales issue. For years US officials have used China's substantial and growing missile deployment in Fujian and Zhejiang provinces as the main reason for US arms sales to Taiwan. As recently as March, 2002, an unnamed senior member of the Bush Administration was reported to have said that a decrease in China's missile deployments would be a precondition for any limit on US arms sales to Taiwan.[36]

Chinese officials, however, expressed frustration at US policymakers who seem to believe that China is now behaving well as a result of the Bush Administration's tougher policy toward China and its clearer support of Taiwan. One Chinese official is quoted as having said, referring to arms sales, "China has been making serious efforts to improve its ties with the US. Anti-terrorism is important to the US, and China's support is important on this front. But you can't expect to request us to support you on counterterrorism and then overlook or even hurt our national security on the other issue."[37] It now appears that China has increased, not decreased, its deployment of missiles aimed at Taiwan from an additional fifty missiles per year to seventy-five, and will soon have as many as six hundred such missiles aimed at Taiwan.

The Bush Administration, of course, is likely to take this response as evidence of the need for a missile defense system to defend Taiwan, but such a step would also present a high level of risk. Indeed, it is clear that arms sales to Taiwan coupled with an arrogant, one-sided approach to bilateral relations have produced a fifty percent increase in the rate at which China has deployed hostile missiles aimed at Taiwan. It seems unlikely that matching these missiles with a missile defense system will produce a reduction in such missiles. More likely, the same pattern will repeat itself, and the deployment of a missile defense system will produce yet another surge in the arms race that is already going on in the region.

Adding to China's concerns is the March 2004 reelection of Taiwan's President Chen Shu-bian, who might want to make Taiwan's de facto independence from China permanent. Beijing has tried to put a positive spin on the election results by highlighting the failure of a referendum on the mainland military threat, which asked whether Taiwan should step up defenses in the face of Chinese missiles pointed at the island. China holds that this is a message to Chen that the Taiwanese public does not really want to "split the motherland."[38] In reality, however, the signals from the vote are mixed. This election produced a higher share of the vote for Chen than the 2000 election had, 50.1 percent compared to 39.0 percent. The defeat of the referendum may be attributed to Taiwanese domestic political tactics. Many voters boycotted it, and the referendum was defeated due to lack of voter participation.[39] No matter how either side interprets the vote, however, Chen has proved in the past to be a pragmatic politician. As McDonald points out, "Chen talked like an independence supporter before his 2000 election, but once in office veered back to the middle after criticism from Taiwan's business elite, which has invested more than $100 billion in the mainland."[40] Given the high level of commitment on the part of Taiwan in the mainland's continued economic development, it may be that China's main concern over its own security comes from the US.

As in the case of the Korean Peninsula, the impact of US missile defense policy in the Straits of Taiwan is difficult to separate from other potentially disruptive inputs. These inputs include US arms sales to Taiwan, the appearance of support for Taiwanese political leaders who advocate independence from China, and increased US military activity in the Pacific. In July of 2004, the US began to deploy up to seven aircraft carrier strike groups in the region as part of a new "Fleet Response Plan." This plan appears to be aimed at increasing the ability of US forces to provide significant combat power in a crisis anywhere in the world. The Chinese, however, do not appear to view it that way. According to Andrew Tam, an expert at the

Institute of Defense and Strategic Studies in Singapore, "It is an unprecedented show of force and a return to gunboat diplomacy. The carrier groups are sent as an affirmation of the US support of the status quo in the Taiwan Straits and the current status of Taiwan."[41] I must conclude, therefore, that US missile defense policy is part of a multidimensional deterrence strategy in the Taiwan Straits. Whether this policy tends to stabilize or destabilize the region depends upon how missile defense fits in the mix of other factors and how it is perceived by the Chinese.

C. Chinese Perceptions of U.S. Missile Defense Policy

Clearly, China would prefer not to contend with a strong US military presence in East Asia. As for the missile defense component of that presence, David M. Finkelstein reports that China's arguments against US missile defense policy can be grouped under two headings: 1) the implications of missile for the viability of China's nuclear deterrent, and 2) the perceived impact of missile defense on the international security environment.[42] With regard to the first of these arguments, the logic of the Chinese position runs about as follows: 1) the US may engage in a first strike since the US does not have a no-first-use policy as does the People's Republic of China (PRC); 2) nearly all of

China's tiny nuclear arsenal would be destroyed by this attack while the rest of it, having been launched in self-defense, would be defeated by the US NMD system; 3) missile defense, therefore, negates China's already minuscule retaliatory capability; 4) conclusion: missile defense represents a survival risk to China.[43]

The argument regarding the perceived impact of US missile defense policy on the international strategic security environment stems from some of the same premises as the first argument. As Finkelstein shows, during the Cold War, the Chinese leadership assumed that the probability of war between the two superpowers was rather slight, and that in any event the prospect that such a war would involve China was slighter still. The overall strategic balance that prevailed under this regime was perceived as advantageous to China, giving it room to pursue economic modernization in the post-Mao era. Now that the US has emerged as the only superpower, China's perception has changed. The one superpower possessing both the "spear" of nuclear weapons and the "shield" of missile defense causes a destabilization in the international order China has previously relied upon for its safety. China, therefore, believes that missile defense will produce a new arms race driven by a desire to regain some degree of balance in the international system.[44] With or without missile defense, that balance must be reached; but the Chinese leadership believes that without defensive US missiles, the balance would be more readily at hand.

Recall my theoretical discussion in section two of this chapter. As I pointed out there, missile defense represents a disruptive input into the international system, which, depending upon the nature of accompanying environmental constraints, can produce either desirable or undesirable effects in the international system. In the US-China relationship, each side assesses these constraints in a different and very one-sided way. Each side has domestic political constraints that must be weighed in the balance. As I have shown in Chapters Two and Three, US missile defense policy has been driven by an array of domestic economic interests as well as factors related to inter-party political competition. China, too, must shore up support for its regime and pursue goals of economic development. The two countries are also strong trading partners, and have become increasingly interdependent over recent years as lower priced Chinese manufactures stock the shelves of countless American discount stores. The complexity of all these factors makes it difficult to assign precise quantitative values to them in calculating the equation that will estimate the impact of missile defense on the East Asian region.

From the US point of view, China has no special right to missile dominance in East Asia, and the missile defense systems now under development will have limited capability against Chinese intercontinental and regional offensive missiles. Consistent with this view, I would argue that US policy on missile defense cooperation in the region should remain firm in support of US alliances while maintaining a constructive, non-confrontational relationship with China. Multi-level dialogue between the US and China as well as moderate rhetoric will be important keys to future stability.[45]

D. Japan and the US-Japan Security Relationship

Deterrence theory teaches us that wars arise from a series of escalating threats and counter-threats, which include expansion in the amounts and types of weaponry available. Military preparedness, which might include missile defense systems, may deter such threats, but may also generate the kind of arms race that will make a challenge to the equilibrium of the system from a hostile external source more likely. I have undertaken this study of US missile defense policy in order to understand more clearly the likely consequences of that policy.

I have found through my brief explorations of the current crisis on the Korean Peninsula, and the tensions between China and Taiwan that US missile defense policy seems more likely to produce escalation of threats and counter-threats than to deter military challenges to the status quo. My recommendation, in contrast to that of the current Bush Administration, is that the US government should dramatically reduce its emphasis on missile defense both for itself and for its allies. This is especially important for East Asia where much is currently at stake, and where miscalculation and misperception could have particularly disastrous consequences. This may mean that US allies in East Asia, such as Japan and South Korea, will need to seek greater independence from US policy in order to regain their balance in these precarious times.

Japan's reaction to NMD must surely be a mixed one. Tokyo cannot avoid being concerned about the North Korean missile threat, or its willingness to carry out its nuclear ambitions. China's growing missile arsenal must also cause worry. If NMD addresses these concerns, it is a valid and useful policy from the Japanese point of view. But if NMD only leads to greater antagonism between the US and China, and causes an arms race that only makes matters worse in East Asia, then Japan has yet another worry. The US-Japan alliance has been the key to East Asian stability and security for more than half a century. Since the alliance rests on the US providing a security umbrella for Japan, it depends heavily on Japanese willingness to trust the wisdom of US security policies. Japan must therefore carefully and critically scrutinize any missile defense program before giving its support.

For example, Japan has wisely questioned whether the confrontational US approach toward North Korea is consistent with Japanese security interests. This skepticism highlights the fact that together the two countries must face an array of challenges ahead. As Michael Krepon points out, "peace in Northeast Asia is an achievement maintained over the past five decades by US military presence and diplomatic flexibility."[46] American willingness to maintain the lines of communication between itself and regional states has been the basis for a half century of stability and prosperity.

The US continues to push Japan to reinterpret or redefine Article 9 of its Constitution to make Japan a fuller partner in East Asian military defense. In the face of the threat from North Korean missile and nuclear weapons programs—and with growing Chinese military power in the background—Japan now seems receptive to constitutional change. Although the current interpretation of the Japanese Constitution poses an obstacle toward forming a clear-cut policy of collective defense, it need not be an obstacle to Japanese participation in a missile defense program. Missile defenses are, by definition, defensive, and therefore present fewer problems for Japan's peace constitution, which rejects the use of force as an instrument of national policy.[47] Furthermore, any other country in the same precarious position Japan now finds itself would seek a nuclear deterrent to offset the various threats it now faces to its security. Since Japan's historical aversion to nuclear weapons makes this alternative unacceptable, some other option must be found. Missile defense may be that option.[48]

In Japan, however, such a policy can proceed only if it has broad support. As Slocombe, Carns, Gansler and Nelson point out, missile defense in Japan already has strong support among those who are more conservative and security-minded; but many traditional left-wing and pacifist groups can never support it. Concern about proliferation and potential new threats to Japan will move many people in Japan to support a more active role for Japan in international security. At the same time, however, broad elements of Japan public opinion remain opposed to a greater Japanese military role. For missile defense to become a serious element in Japanese military programs, therefore, it will be necessary to win the support of many previously uncommitted groups.[49]

Therefore, I suggest that there is a positive correlation

between US-Japan cooperation on missile defense and the importance of careful diplomacy by the US in dealing with Japan. I agree that as cooperation between the US and Japan on missile defense continues to go forward, the US must take care to avoid actions that would be interpreted as placing too much pressure on the political decision making process in Japan. Japan must be allowed to make up its own mind on missile defense, taking into account its own relationships with other countries in the region. Excessive US pressure on Japan would be counterproductive in terms of further missile defense collaboration and could harm the US-Japan security relationship. If the US wants to be able to rely upon Japanese support in facing a range of uncertain threats, it must take Japanese needs into account.

E. General Assessment of Missile Defense Policy Impact on East Asian Security

Our brief look at East Asian security dilemmas shows that the impact of US missile defense policy in the region is difficult to measure. On the Korean Peninsula, it is bound up with a range of countervailing forces, such as the North Korean push to develop a nuclear arsenal on the one hand versus the desire for reunification of the two Koreas on the other. Here inflammatory US rhetoric and equally defiant

North Korean responses have done as much to put solutions to these problems out of reach as the introduction of missile defense into the region has done. Yet, it is clear that fully deployed missile defenses would not reduce tension on the Peninsula.

Similarly, in the Taiwan Straits, the impact of US missile defense policy cannot be isolated from such other factors as US arms sales to Taiwan or China's desire to treat Taiwan as a rebellious province that must one day return to the arms of the motherland. Again, however, as is the case with the two Koreas, US missile defense policy appears to increase rather than decrease tension.

When we turn our attention to perceptions of US missile defense policy from the points of view of the two most important regional powers, China and Japan, we find no easy or simple answers. China sees US missile defense policy as a threat to its survival interests while Japan finds it a challenge to its longstanding desire to remain at least one step removed from military action. How the US deals with these concerns diplomatically will largely determine whether the impact of US missile defense in East Asia is positive or negative in the long run.

The complexity of these issues is enough by itself to recommend caution in the implementation of US missile defense policy in East Asia. There is, however, another dimension to this complexity, which weighs even more heavily on the side of caution in implementing this policy. This dimension is the war against terror, which the current Bush Administration has linked to its missile defense policy since the terrorist attacks on New York City and Washington, D.C. in September, 2001. In the next section of this chapter, I will show that the US war against terror operates in conjunction with US missile defense policy to create additional concerns for East Asian security.

4. Indirect Effects on East Asian Security: Missile Defense and the War against Terror
A. The Link between Missile Defense and the War against Terror

Shortly after the September 11, 2001 terrorist attacks in the US , the Bush Administration began to tie together the war on terror and its missile defense policy. The purpose of this was to present the view that this war has very broad dimentions,which support the need for missile defense. There are three steps in the Bush Administration argument that leads to this conclusion. First, the current Bush Administration holds that the enemy in this war includes not only non-governmental terrorist organizations like Al Qaeda, which brought down the twin towers of the

World Trade Center in New York City, but also governments that harbor, encourage, fund, or in any way give such organizations aid and comfort. Speaking at the United Nations on November 10, 2001, the President said:

> "Every nation has a stake in this cause. As we meet, the terrorists are planning more murder ...perhaps in my country, or perhaps in yours. They kill because they aspire to dominate. They seek to overthrow governments and destabilize entire regions."[50]

The same broad portrayal of the enemy applies to US military operations in Iraq. This is demonstrated by President Bush's remarks on October 7, 2002 linking the government of Iraq to Al Qaeda, stating: "We know that Iraq and Al Qaeda have had high level contacts that go back a decade, and that Al Qaeda leaders who fled Afghanistan went to Iraq."[51] He went on to say, "Alliance with terrorists could allow the Iraqi regime to attack America without leaving any fingerprints."[52]

Secondly, since the attacks of September 11, 2001, the US government has held the position that the war on terrorism is a global conflict. US Secretary of State Colin L. Powell has stated that, "no country has the luxury of remaining on the sidelines. There are no sidelines."[53] In response to this appeal, Japan, Australia and the parties to the ASEAN treaty were among the nations who pledged to support the US in various ways to oppose terrorism in the days after the September 11 attack. Within one week of the attacks, the Japanese government under Prime Minister Koizumi put together a seven-point package of actions to support the US. These steps included measures involving Japan's Self-Defense Forces, expanded security for US forces and facilities in Japan, humanitarian assistance to affected countries and displaced persons, measures to support the world economy, and strengthened international cooperation in sharing information and immigration control.[54]

Finally, the current Bush Administration has linked the war against terror directly to its missile defense policy. President Bush expressed this view in a statement on December 17, 2002:

"I have made clear that the United States will take every necessary measure to protect our citizens against what is perhaps the gravest danger of all: the catastrophic harm that may result from hostile states or terrorist groups armed with weapons of mass destruction and the means to deliver them. Missile defenses have an important role to play in this effort. We have adopted a new concept of deterrence that recognizes that missile defenses will add to our ability to deter those who may contemplate attacking us with missiles."[55]

From the US point of view, therefore, terrorism threatens the entire security community, including Japan, and jeopardizes all regions of the world, including East Asia. Consistent with this point of view, it becomes all the more necessary not only to move forward as quickly as possible on the deployment of missile defenses, but also to secure the support of US allies for the military buildup, including those in East Asia.

B. Is Missile Defense the Right Weapon against Terror?

The zeal of the current Bush Administration for missile defense, however, may be unwarranted. Before US allies jump on the missile defense bandwagon, they must ask whether missile defenses of any kind, even if they had been fully operational, would have deterred the terrorist attacks of September 11, 2001, or any other such attacks whether before or after that date. So far, missiles have not been used as offensive weapons by terrorist organizations. Nor has there been any use of missiles against US and coalition forces in the 2003-2004 military operations in Iraq, which the current Bush Administration considers to be a part of its war against terrorism. At the present time, the US has no missile defense system in place capable of serving as a deterrent to such attacks. This would lead any fair-minded observer to doubt the credibility of the claim that such defenses are necessary.

Such skepticism, however, does not seem to lessen the desire of President Bush for building these defenses. To gain an understanding of President Bush's desire on this point I speculate that throughout the current crisis, the Bush Administration has been determined to avoid what it has viewed as the mistakes of the past in foreign and military policy. In the Persian Gulf War of 1991, a US-led coalition routed Iraqi forces in the field and drove them out of Kuwait, which they (the Iraqis) had invaded. Iraqi offensive missiles, the hastily constructed Scuds, were shown to be largely ineffective, despite one rather devastating attack against a US military barracks in Riyadh, Saudi Arabia. At the same time, however, the missile system deployed by the US to defend against the Scuds during the war also proved to be almost totally ineffective. This missile defense system, the Patriot system, which had originally been designed to defend against aircraft, had trouble locating the Scud warheads amid the debris of the poorly made Scuds, which often broke up in flight. This experience demonstrated that missile defense at that time was an ineffective tactic.

Instead of abandoning the missile defense concept, however, the Pentagon continued its research and development effort throughout the years of the Clinton Administration. As we have seen in the previous chapters, political support for missile defense remained firm in Congress, and eventually prevailed in the executive branch as well with the victory of George W. Bush in the 2000 election. At the time of the terrorist attacks of September 11, 2001, Congress was at work on the fiscal 2003 defense budget. This was another moment when the missile defense concept might well have been abandoned. Instead, the President sought to forge the link between missile defense and the war on terror. With political support in Congress still firm, funding for missile defense in the fiscal 2003 US defense budget received a considerable boost, and has increased in both of the subsequent two fiscal years.

Missile defense has not been without opposition in Congress during this period. As I pointed out in the previous chapter the Democrats held a slender majority in the Senate during 2001 and 2002, and were in a position to question whether such large budgets for missile defense would shortchange anti-terrorism efforts.[56] As then-Majority Leader Senator Thomas A. Daschle put it, "How could anyone think we are more likely to be the target of a ballistic missile attack than another terrorist incident?"[57] Nevertheless, the President was able to persuade Congress to appropriate almost every dollar he requested for missile defense during these years, and this encourages him to believe that US allies will also support his missile defense policy.

The President's success in gaining funding for missile defense is all the more remarkable in view of the difficulties it has experienced in testing. As Bradley Graham has reported, the PAC-3 missile defense system, the one now favored by the Pentagon, has recently come out of an eight-year research and development program in which it experienced delays and cost overruns. The PAC-3 missile system was designed to replace the Patriot missiles that had been ineffective against the Scuds in the Persian Gulf War. Unlike the older Patriot missiles that destroyed their targets by blowing up near them and blasting them out of the sky, the PAC-3 interceptors are built without explosives in them, and instead knock out offensive warheads by colliding with them. Military planners think this approach, known as "hit-to-kill," is more reliable against nuclear, biological or chemical warheads than defensive missiles that use explosives to destroy their targets. The problem seems to be that so far the system has not proven itself in any realistic sort of test.[58]

In the period prior to the 2003 war in Iraq, the Pentagon renewed its commitment to the PAC-3 program. Despite test failures, which might well have added a yearlong delay in any effort to accelerate production on the missile system, aides to Defense Secretary Donald Rumsfeld decided to increase production without waiting for any further tests. At that time, the US Army had only thirty-eight PAC-3 missiles in its inventory, and expected only another fifteen by the end of 2003.

At the time the Iraq war broke out, US military analysts believed that Iraq possessed perhaps twelve to twenty-five mobile Scud missiles. Previously, the US Army kept two Patriot batteries in the Persian Gulf region guarding US military facilities—but none equipped with PAC-3 interceptors. Despite all of the technological problems experienced by the PAC-3 system, President Bush ordered that additional Patriot batteries armed with available PAC-3s be deployed in the event Scuds were launched against US forces.[59]

As it turned out, no defensive missiles were necessary. The Iraqis were unable to launch even one offensive missile. This was not, however, due to the deterrent effect of the PAC-3 missile defenses. The Iraqis either did not have time or did not have an adequate plan for the use of their own missiles in case of attack. In either case, I do not now have current data to show the preparedness or effectiveness of PAC-3 missile defenses in the theater of war.

Recent tests of missile defenses deployed in Alaska, however, have shown mixed results. On December 15, 2004, an interceptor missile failed to launch in a flight test of the system,[60] and a similar failure occurred on February 13, 2005.[61] The Aborted tests cast fresh doubt over when Defense Secretary Rumsfeld would put the system on alert, and brought about renewed criticism of the failure to adequately test the system before deployment.[62] On February 24, 2005, however, the Missile Defense Agency announced a successful short range test in which a missile launched from a cruiser in the Pacific Ocean destroyed a missile launched from Hawaii.[63] Nevertheless, a successful test at such short range indicates that the system has a long way to go before it is able to defend an all out missile attack on the US.

5. Conclusion: Caution is the Watchword

The pluralistic security community that now embraces the US, Western Europe and Japan reflects the profound reality that we now exist in the unique position in which war among the most developed countries of the world is unthinkable. Yet, that community is threatened from the outside by terrorist groups, and potentially dangerous regional conflicts. My focus in this chapter has been on the impact of US missile defense policy on the various security dilemmas now facing East Asia. I have specifically examined the two most volatile conflict areas in the region, the Korean Peninsula and the Taiwan Straits. I also looked at the two most important regional actors, China and Japan. I wanted to know how US missile defense policy would affect these areas of conflict, and how this policy would be viewed by those two important regional actors.

My general conclusion is that if the US implements its missile defense policy in East Asia in a hasty and forceful manner, it is a policy that is not likely to preserve a regime of general deterrence in East Asia. If imposed upon East Asian allies of the US in such a manner, missile defense will tend to disrupt rather than reinforce the status quo. This will require a perilous and temporary regime of immediate deterrence in which the forces available to resolve conflict peaceably are difficult to measure and control.

Although I emphasize caution in the implementation of US missile defense policy in East Asia, I cannot entirely rule out a possible role for missile defense in establishing a successful regime of general deterrence in East Asia at some time in the future. I do not recommend halting all research and development efforts in missile defense, nor should diplomatic efforts to acquaint US allies with the potential utility of missile defense be halted. A realistic assessment of the need for missile defense in East Asia permits a reasonable amount of time to prepare US allies in the region to incorporate missile defense in their strategic planning.

Given the indirect impact of global terrorism on regional conflicts, more resources should be devoted to the development of effective countermeasures against terrorism within each nation of the security community, including Japan. In other words, a greater emphasis should be placed on homeland security. This is more likely to involve international sharing of resources among the nations of the community, not only in the form of shared intelligence but also in the form of stepped up training in counter-terrorism tactics.

In a larger context, however, the external threats to the security community in general and East Asia in particular must be handled through firm, straightforward negotiations. No amount of military superiority can guarantee the security of every US ally unless the US is also willing to take into account their diverse interests and problems.

Notes

1 Robert Jervis, "Theories of War in an Era of Leading-Power Peace" (Presidential Address, American Political Science Association, 2001) *American Political Science Review*, vol. 96, no.1, March, 2002, p. 1.
2 Ibid.
3 Ibid.
4 Ibid.
5 Futoshi Shibayama, "Japan-US Cooperation and the Road to Alliance Missile Defense (AMD)," *The Journal of International Security*, vol. 29, no. 4 (March, 2002) p. 105.
6 Paul Huth and Bruce Russett, "General Deterrence between Enduring Rivals: Testing Three Competing Models," *American Political Science Review*, vol. 87, no. 1 (March, 1993) p. 61.
7 Ibid., p. 62.
8 Ibid., p. 63.
9 Kristen Eichensehr, "Broken Promises: North Korea's Waiting Game," *Harvard International Review*, vol. 23, no. 3 (Fall, 2002) p. 1.
10 Huth and Russett, op.cit., p. 62.
11 Robert C. North, *War, Peace and Survival* (Boulder, Colorado: Westview Press, 1990) p. 160.
12 Richard N. Rosecrance, *Action and Reaction in World Politics: International Systems in Perspective* (Boston: Little, Brown and Co., 1963) p. 229.
13 Michael R. Gordon, "US Toughens Terms for North Korea Talks," *New York Times*, July 3, 2001, p. A7.
14 Kenneth Lieberthal, "The United States and Asia in 2001: Changing Agendas," *Asian Survey*, vol. xlii, no. 1, January/February 2002, p. 1.
15 Gordon, op.cit., p. A7.
16 Doug Struck, "US Plays Down Talks with North Korean Officials," *Washington Post*, October 6, 2002, p. 29.
17 The White House, "President Focuses on US Economy, Iraq and North Korea," www.whitehouse.gov/news/releases/2003/01/2003/0102.html, January 2, 2003, p. 1.
18 Reuters News Agency, "North Korea Tells US to Mind Its Own Business," *Washington Post*, June 16, 2003, p. A01.
19 Ibid.
20 Selig S. Harrison, "The Nuclear Crisis on the Korean Peninsula: Avoiding the Road to Perdition," *Current History*, April, 2003, p. 152.

21 Ibid.
22 Ibid., p. 153.
23 Ibid., p. 165.
24 "Pyongyang said Ready to End Nuclear Quest," *Japan Times*, February 24, 2004, p. 1.
25 "North Korea talks end without major progress," *Japan Times*, June 27, 2004, p. 1.
26 "North Korea rejects US nuclear offer," *Japan Times*, June 29, 2004, p. 4.
27 "North Korea's 'Anybody but Bush' Stalling Policy won't do Kim any good," *Japan Times*, March 6, 2004, p. 1.
28 Joe McDonald, "China Call US to Start North Korea Talks," *Associated Press*, March 5, 2005, p. 1.
29 Joe McDonald, "China: North Korea Willing to Discuss Nukes," *Associated Press*, March 6, 2005, p. 1.
30 Edward Cody, "China Sends Warning to Taiwan with Anti-Secession Law," *Washington Post*, March 8, 2005, p. A12.
31 Alan D. Romberg and Michael McDevitt (eds.) *China and Missile Defense: Managing US-PRC Relations*, (Washington, D.C.: The Henry L. Stimson Center, 2003) p. 23.
32 James Melvenon, "Missile Defenses and the Taiwan Scenario," in Alan D. Romberg and Michael McDevitt (eds.) *China and Missile Defense: Managing US-PRC Relations* (Washington, D.C.: The Henry L. Stimson Center, 2003) p. 54.
33 Ibid., p. 52.
34 Dali L. Yang, "China in 2001: Economic Liberalization and Its Political Discontents," *Asian Survey*, vol. XLII, no. 1, January/February, 2002, p. 15.
35 John Pomfret, "China Suggests Missile Buildup Linked to Arms Sales to Taiwan," *Washington Post*, December 10, 2002, p. A01.
36 Ibid.
37 Ibid.
38 Joe McDonald, "China left guessing how Chen will handle mandate," *Japan Times*, March 26, 2004, p. 21.
39 Ibid.
40 Ibid.
41 "China cautions US regarding Taiwan," *Japan Times*, July 10, 2004, p. 3.
42 David M. Finkelstein, "National Missile Defense and China's Current Security Perceptions," in Alan D. Romberg and Michael McDevitt (eds.) *China and Missile Defense: Managing US-PRC Relations* (Washington, D.C.: The

Henry L. Stimson Center, 2003) p. 40.
43 Ibid., p. 41.
44 Ibid., pp. 42-43.
45 Walter B. Slocombe, Michael P. C. Cairns, Jacques S. Gansler, and C. Richard Nelson, *Missile Defense in Asia*, (Washington, D.C.: The Atlantic Council of the United States, 2003) p. xii.
46 Michael Krepon, "Missile Defense and Asian Security," in Alan D. Romberg and Michael McDevitt (eds.) *China and Missile Defense: Managing US-PRC Relations* (Washington, D.C.: The Henry L. Stimson Center, 2003) p. 73.
47 Slocombe, et al., op.cit., p. 7.
48 Ibid.
49 Ibid., p. viii.
50 The White House, "President Bush Speaks to United Nations," http://www.whitehouse.gov/news/releases/2001/11/20011110-3.html, November 10, 2001, p. 1.
51 The White House, "President Bush Outlines Iraqi Threat," www.whitehouse.gov/news/releases/2002/10/20021007-8.html, October 7, 2002, p. 1.
52 Ibid.
53 Secretary of State Colin L. Powell, *Patterns of Global Terrorism, 2001* (Washington, D.C.: US Department of State, 2002) p. 5.
54 Ari Fleischer, "President Welcomes Japan's Support," Statement by the White House Press Secretary, September 20, 2001.
55 The White House, "President Announces Progress in Missile Defense Capabilities," www.whitehous.gov/news/releases/2002/12/20021217.html, December 17, 2002, p. 1.
56 Helen Dewar, "Missile Defense Funding Increased," *Washington Post*, June 27, 2002, p. 9.
57 Ibid.
58 Ibid.
59 Ibid.
60 Bradley Graham, "US Missile Test Fails," *Washington Post*, December 16, 2004, p. A5.
61 Rick Lehner, "Missile Defense Flight Test Conducted," *Missile Defense Agency News Release*, February 14, 2005, p. 1.
62 Graham, op.cit., p. A6.
63 Rick Lehner, "Aegis Ballistic Missile Test Flight Successful," *Missile Defense Agency News Release*, February 24, 2005, p. 1.

Chapter V

Conclusion

1. Review of Findings

This research has had two basic purposes. The first, which I accomplished in my second and third chapters, was to explore how US missile defense policy is made. I began this exploration in Chapter Two by surveying the history of missile defense policy in an effort to determine the nature of the structures and processes that have guided the development of this policy since its inception. I wanted to know whether the system that produced US missile defense policy was an open or a closed one, and to identify the range of actors who have exercised influence over the implementation of that policy. Also in Chapter Two, I sought evidence regarding the degree to which the material interests of those actors have been served in missile defense policymaking. Finally in that chapter, I tried to judge in a broad outline which sets of actors would exert control over missile defense policy.

I found that over time the answers to these questions could change, and that missile defense policy making has often been in a state of flux. I noticed that as material stakes in missile defense policy have increased, and as budgets for missile defense have grown, controversy over the policy has also increased. With both material stakes and controversy on the increase, more and more interested actors have attempted to enter and influence the process. As this political pressure continued to increase, missile defense advocates moved to protect its further development from adverse political forces. The key event in this process was President Reagan's announcement of the Strategic Defense Initiative (SDI) in 1983, and his creation of the Strategic Defense Initiative Organization (SDIO) within the Pentagon. Since then, missile defense policy has become entrenched in a bureaucracy of its own. During the Clinton Administration it was called the Ballistic Missile Defense Organization (BMDO), and it was renamed the Missile Defense Agency (MDA) during the administration of President George W. Bush. It now enjoys the favorable attention of a Republican president and Republican majorities in both houses of Congress.

Given this foothold in the Pentagon bureaucracy, US missile defense policy could have survived the winds of controversy that had begun to swirl around it during the Administration of the first President Bush. The various criticisms of that policy have changed little over its history: missile defense is too expensive; it is technologically unfeasible; it encourages an arms race in offensive weapons; and it disrupts various alliance relationships. As valid as all of these criticisms have been from the beginning, they have had little impact against an entrenched bureaucratic position. Added to this was the fact that some members of Congress have found that they have benefited considerably from federal government spending on missile defense projects, especially in the form of the extensive research and development programs these projects have generated. Furthermore, these projects have had popular support in congressional districts where they have produced jobs for voters. Despite the end of the Cold War in the early 1990's, this support has caused members of Congress, motivated by the goal of reelection, to advocate more spending on missile defense.

In 1993, the Republican Party lost control of the presidency, but in 1995 they gained control of Congress. The resulting change in the relationship between president and Congress led us to continue my exploration of missile defense policy from another vantage point in Chapter Three. As I noted there, presidents often have difficulty persuading the members of Congress to comply with their wishes and support their policies. For the president, the political environment of policy decisions is crucial. Having a majority in Congress of the president's own party that generally support the president's policy goals helps considerably; but it is also important for a president to use the informal tools of presidential power. Specifically, he must move quickly on his goals, and communicate them clearly to Congress. As we saw in our comparison of Presidents Bill Clinton and George W. Bush, President Clinton failed to move quickly enough while he had a congressional majority in the first two years of his first term. As a result he not only missed a number of opportunities to achieve policy goals, such as in the area of health care, but his party also lost control of Congress in the first national election after his inauguration. This forced President Clinton into a reactive position with respect to Congress at a time when he wanted to re-order the nation's missile defense priorities toward a greater emphasis on the theater missile defense (TMD) rather than national missile defense (NMD). The Republican Congress, however, invariably sought to increase the President's budget proposals for missile defense so that more money could be put into NMD.

The built-in institutional antagonism between president and Congress in the American governmental system calls upon presidents to be especially persuasive in gaining congressional support. Even members of the president's own party in both houses of Congress support the president only about seventy percent of the time on key issues where the president has taken a stand. But where the president has provided a range of incentives to key figures in Congress, such as the leading members of the Senate and House Armed Services Committees, and where these incentives are relevant to the members' goal of achieving reelection to office, the president can often count on support from those key figures. The second President Bush was able to do this more effectively than President Clinton did, and this support translated into higher appropriations for the president's policy goal, which in this case was NMD.

I also found that pro-missile defense members of Congress were particularly effective at using missile defense in the performance of the various activities that have always led to re-election: advertising, position-taking and credit claiming. The Rumsfeld Commission, called into existence by Congress to study the missile threat, helped place missile defense on the agenda at a time when so-called "rogue" nations (North Korea, Libya, and Iran were the ones most often mentioned) appeared to be gaining ground in the development of offensive missile capabilities. This set the stage for several legislative initiatives including the Theater Missile Defense Improvement Act of 1998 and the National Missile Defense Act of 1999. The purpose of both of these laws was to establish a basic policy in favor of missile defense so that those who supported the legislation could claim to be "strong on defense of America's vital interests." Both bills passed rather easily, indicating the degree to which the political environment favored missile defense at this time.

Another finding with respect to Congress concerned the apparent importance of campaign contributions from defense contractors and the presence of defense contractors as major employers of the constituents of the members of Congress who occupied key positions on the House Armed Services Committee. This seemed to be a key factor associated with support for missile defense. Finally, I noted the apparent lack of discussion throughout the various debates over the Theater Missile Defense Improvement Act of the potential impact of these weapons systems on the interests of American allies. Subsequent to the passage of this legislation, Japan entered into a bilateral collaboration with the US in the research and development of TMD technology. Although Japan agreed to be a party to this effort, its cooperation may have been more a matter of demonstrating trust in its ally than of eagerness to develop a new weapons system. The possible destabilizing effect of

TMD on the balance of power in East Asia was not part of the debate among US policymakers, nor was the danger that TMD development could stimulate greater production of offensive weapons among potential adversaries in the region. The even greater danger was that Japan would compromise its own security interests to join the US in a policy that was largely the product of competing political interests in the US.

The possibility of such a danger brought us to look at the impact of US missile defense policy on East Asian security. In Chapter Four, I noted the heightened fear currently existing in China that US missile defense policy threatens the survival interest of China, and that China's response to US missile defense deployment would necessarily be to increase its production of offensive weapons. This is taking place at a time of increasing tension over the relationship between China and Taiwan and speculation about US action in case the conflict erupts into violence.

Similarly, I noted that confrontational gestures by the US toward North Korea have stalled progress toward reunification on the Korean Peninsula, and have triggered a confrontational response from the North Koreans. This response has taken the form of a re-initiation of the North Korean nuclear weapons program, which in turn has led to several rounds of talks among the six nations most affected by the situation (North Korea, South Korea, China, Japan, Russia and the US) to defuse the crisis.

In sum my research suggests that the political environment within the US determines policy that in turn affects the international political environment. US missile defense policy is an illustration of this phenomenon. Furthermore, the effect of this policy on the international environment appears to be destabilizing. I may suppose that as the international environment grows more unstable, attitudes toward US leadership of the international system will grow more skeptical and less supportive unless some corrective measures are taken by the US.

2. Further Considerations

This research offers insight into the dynamics of domestic policymaking and their impact on policies that have international implications. As this research suggests, US missile defense policy may be contributing to unfavorable attitudes toward US international leadership. This impact may be far greater than most US leaders are willing to acknowledge. As Richard Butler, the former Australian Ambassador to the United Nations (UN) who led the negotiating team for the disarmament of Iraq from 1997 to

1999, has noted: US missile defense policy will have "grave consequences" for the international arms control and non-proliferation regimes."[1] Butler correctly emphasizes that the rationale for this policy has changed in an important way, one which I have also documented in this book. During the Cold War era, the intended purpose of missile defense was to neutralize a massive nuclear attack from a competitor in the international system, the USSR. The current rationale, however, stresses the irrationality and unpredictability of a range of potential foreign adversaries, i.e. the so-called "rogue" states. Although it is conceded that none of these states presently possesses long-range ballistic missile capabilities, pro-missile defense forces in the US continue to argue that those states may, indeed, will acquire such capabilities before long.[2] These forces further argue these rogue states cannot be bound to adhere to treaty obligations they have entered into simply because of their desire to harm the US, and that existing treaty and control mechanisms are unverifiable.

Butler responds first that although there is evidence that potential regional adversaries around the globe are at work to gain greater weapons capabilities, there is little or no evidence to suggest that they do not honor their treaty obligations. As to the argument regarding the verifiability of treaty and control mechanisms, Butler points out that this argument "confuses verification with enforcement," and that "the activities and intentions of such states can be verified to a very high degree of accuracy, if sufficient effort by international and national agencies is employed to that end."[3] Enforcement is a more difficult matter, but this too can be accomplished if the major powers, i.e. the members of the pluralistic security community, "determined that they were prepared to enter into enforcement action, both politically and militarily."[4]

The most important argument against US missile defense policy, however, is its impact on other nuclear weapons states. President Bush and other missile defense advocates within his administration have asserted that the purpose of missile defense is strictly defensive and that its intent is not only to shield the US but also its allies, and that other nuclear powers should be assured that the missile shield will have no impact on their security interests. As I have shown in Chapter IV, China is one nuclear state that views US missile defense as a threat to its very survival. As Butler points out further, other nuclear states, such as Russia, may be even more skeptical if it appears to them that the true goal of missile defense is to gain US military dominance of the upper atmosphere of the earth and outer space.[5] Military planners in Russia, China and elsewhere must calculate the impact the US defensive shield would have on their ability to deter a nuclear attack upon them, which would be launched from the US. The only possible response to this threat would be to build large numbers of offensive missiles and to invest heavily in other forms of military production until they reach a point where these countries feel reasonably certain that they are able to deter US aggression.

This is hardly a satisfactory result for East Asia or any other region of the world. The security community, in which war is unthinkable, would disappear to be replaced once again by some kind of balance of power system characterized by cold war and nuclear standoff. All of this cannot be blamed entirely on US missile defense policy, nor can it be said that this policy would be the cause of this disastrous state of affairs. Rather, the most likely explanation would be that missile defense is symptomatic of a larger problem. My analysis suggests that one source of this problem is at the center of American domestic politics where we find intense partisan competition, institutional fragmentation, and a system in which the stability of the international system falls far from the top of the political agenda.

3. Recommendations

My research suggests a number of recommendations.

First, I believe that a greater knowledge and awareness of international security issues needs to be attained both within the American government and the American electorate. At the present time in the US, the links between weapons for national defense and the impact of those weapons on perceptions of international security and stability within other countries are not well understood either by governmental leaders or by American voters. I believe that a broad-based informational campaign should be mounted within the US to inform both political leaders and voters about these perceptions.

Secondly, I believe there is a lack of coordination within the policymaking institutions of American government regarding international security issues and the impact of national defense decisions on those issues. Decisions that affect international security should not be handled in quite the same way as other decisions involving budgets or legislation. The Foreign Relations Committees of the two houses of Congress must be prepared to examine Pentagon budget proposals, and hold hearings on the international security implications of weapons system deployments. There should also be an office within the National Security Council that provides a fact finding service for both Congress and the White House on these issues. The State Department should also be more proactive in supplying information to both Congress and the President on the implications for international security of US weapons procurements.

Thirdly, the US government needs to engage in more frequent and deeper communications both with allies and potential adversaries regarding weapons procurement and deployment decisions. Implicit in my findings is the fact that there was little transition between the policies of the Clinton Administration and those of the second Bush Administration. President George W. Bush seemed to want to pursue a foreign policy as different from that of his predecessor as possible, and to do so without transition. If the US is to lead the security community and act to maintain peace in the international community, then there must be more continuity and less abrupt change from one administration to another. Although many in the international community today might welcome an abrupt and complete turnabout in American foreign policy, such a great change that affects the lives of so many should occur at a more measured pace.

Finally, as a recommendation to US allies, I suggest that one of the lessons of US missile defense policy making since 1983 is that the US can sometimes be an unreliable and unpredictable ally. Missile defense began as a Cold War strategy; but when the Cold War ended, the effort to produce a missile defense system did not end. It simply acquired a new rationale. If the US does not make changes in the way in which it coordinates its diplomatic efforts and missile defense policy with a careful eye toward the interests of both its allies and its adversaries, then its allies must proceed along a more independent path of their own.

4. Suggestions for Further Research

I believe that more systematic efforts need to be made to explore linkages between the policymaking processes within the institutions of American government and the impact of policy on international relations. My explorations have turned up a number of possible future directions for research. A closer look at the committees of Congress that handle foreign relations and defense-related issues would be most useful. A more systematic study of the links between the Missile Defense Agency and the defense contractors who benefit from its growing budgets would also deepen our knowledge. Similarly, a closer study of the policy preferences of the presidents would be very much in order.

Finally, I think that further study of the same links I have been exploring in this book between domestic political processes and their impacts on international relations would advance the study of peace research.

Notes
1 Richard Butler, *Fatal Choice: Nuclear Weapons and the Illusion of Arms Control* (Boulder, Colorado: Westview Press, 2001), p. 109.
2 Ibid., p. 104.
3 Ibid., p. 105.
4 Ibid.
5 Ibid., p. 107.

Index

A

Abercrombie, Neil, 148
ABM (Anti-Ballistic Missile) Treaty, 20, 41, 53, 54, 56, 57, 58, 60, 79, 80, 82, 84, 85, 89, 97, 108, 132, 134
Advertising, 105
Afghanistan, 20, 121, 162
Agnew, Spiro, 57
Allard, Wayne, 129
Allen, Thomas, 148
Al-Qaeda, 20, 162, 198, 199
Anderson, B. James, 42, 50, 52
Asia, 21
Aspin, Les, 64, 65, 66, 81, 109

B

Barone, Michael, 135
Bateman, Herbert, 148
Bethe, Hans, 57
Blagoyevich, Rod, 148
BMD (Ballistic Missile Defense), 111, 112, 113
BMDO (Ballistic Missile Defense Organization), 22, 66, 70, 73, 75, 84, 90, 91, 92, 119, 121
Boeing Corporation, 46
BUR (Bottom Up Review), 64
Bush (Jr.) Administration, 36, 37, 93, 100, 110, 113, 118, 121, 136, 174, 178, 179, 182, 185, 186, 193, 198, 201, 225

Bush (Sr.) Administration, 109, 110, 116, 137, 151
Bush, W. George, 15, 16, 20, 31, 32, 41, 62, 76, 77, 78, 84, 86, 90, 93, 97, 98, 99, 107, 108, 118, 119, 120, 121, 122, 124, 136, 137, 151, 152, 153, 173, 174, 175, 177, 181, 184, 185,199, 202, 203, 214, 215, 216, 217, 222, 225
Butler, Richard, 220, 221, 222

C

Canada, 71
Chambliss, Saxby, 148
Chen, Shu-bian, 187, 188
Cheney, Dick, 136
China, 18, 35, 125, 127, 140, 159, 162, 169, 174, 178, 179, 182, 183, 184, 185, 186, 187, 188, 189, 190, 191, 192, 193
CIA (Central Intelligence Agency), 70
Clinton Administration, 68, 70, 73, 78, 81, 83, 85, 91, 92, 109, 110, 112, 114, 116, 117, 132, 141, 172, 173, 183, 184
Clinton, Bill, 16, 31, 32, 41, 64, 68, 74, 75, 81, 84, 85, 92, 98, 99, 107, 108, 114, 115, 116, 118, 119, 123, 124, 125, 132, 134, 137, 151, 172, 173, 174, 185, 216, 217
Cochran, Thad, 130
Cohen, Jeffrey and Nice, David, 32
Congress, 16, 17, 19, 21, 22, 23, 25,

230 Index

26, 27, 31, 32, 33, 34, 41, 43, 45, 46, 47, 48, 49, 57, 58, 60, 61, 63, 71, 72, 73, 77, 78, 81, 84, 85, 86, 88, 92, 94, 97, 98, 99, 100, 101, 102, 103, 104, 105, 106, 107, 108, 109, 110, 114, 115, 116, 117, 118, 119, 121, 123, 124, 125, 126, 128, 129, 130, 131, 133, 134, 135, 136, 138, 139, 141,143, 145, 146, 147, 152, 158, 200, 203, 204, 214, 215, 216, 217, 218, 224, 225, 226
Cold War, 18, 58, 64, 65, 85, 150, 161, 190, 215, 221
Cooper, F. Henry, 62
Coyle, Philip, 75
Credit claiming, 106, 135

D

Daeschle, Tom, 136, 203
Democratic Party, 110, 115
Détente, 56
Department of Defense (Pentagon), 15, 22, 23, 25, 29, 46, 47, 55, 58, 59, 60, 62, 63, 69, 73, 75, 80, 81, 82, 83, 86, 90, 108, 129, 141, 144, 159, 200, 201
DIA (Defense Intelligence Agency), 70
Dole, Bob, 125

E

East Asia, 18, 23, 35, 36, 37, 67, 118, 156, 157, 158, 161, 162, 164, 175, 186, 189, 190, 192, 193, 197, 201, 207, 208, 219, 223
East Asian Security, 34, 156, 177,

193, 196, 198
Edward, George, and Wayne, Stephen, 124
EU (European Union), 178
Europe, 21, 37
Extended deterrence, 164

F

FAS (Federation of American Scientists), 57
Finkelstein, David, 189, 190
Fleet Response Plan, 188
Fletcher, James, 59
Foreign Relations Committee, 224
Fort Bliss, 146
Fort Greeley, 78, 153

G

GAO (General Accounting Office), 63, 91, 153
General deterrence, 163, 164, 165, 166, 167, 170, 207
Gibbons, Jim, 148
Grand Forks, 58
Glassboro Summit, 56, 82
Gore, Al, 137
Graham, Bradley, 30, 131, 204
Greenland, 120
Grenada, 32

H

Heclo, High, 42, 48
Hefley, Joel, 148
Hilleary, Van, 148
Hoffman, Fred, 59

Hoffman, Stanley, 34
Hostettler, John, 148
House Armed Services Committee, 45, 46, 64, 73, 217, 218
House Defense Appropriations Subcommittee, 15
House International Relations Committee, 145
House National Security Committee, 144, 146
House (of Representatives), 68, 77, 92, 103, 106, 109, 122, 144
Hussein, Saddam, 35, 122, 162

I

IAEA (International Atomic Energy Agency), 164, 176
ICBM (Inter-Continental Ballistic Missile), 29, 54, 184
Immediate deterrence, 153, 164, 166, 167
India, 131
Inouye, Daniel, 130
Iran, 19, 125, 126, 131, 160, 170, 171, 217
Iraq, 19, 62, 126, 150, 162, 199, 201, 205
Iron triangle, 30, 42, 43, 44, 45, 46, 47, 51, 52, 55, 82
Israel, 171
Issue network, 30, 42, 47, 48, 49, 50, 51, 52

J

Jakarta, 180
Japan, 37, 159, 161, 164, 166, 171, 176, 179, 192, 193, 194, 195, 196, 197, 200, 201, 206, 207, 208, 218, 220
Jeffords, James, 136
Jervis, Robert, 160, 161
Jiang, Zemin, 185
Johnson Administration, 55, 56, 82
Johnson, Lyndon, 55, 56, 82
Jones, Walter, 148
JSDF (Japan's Self-Defense Forces), 175, 200

K

Kasich, John, 148
Kelly, James, 175
Kennedy, Patrick, 148
Kerry, Bob, 133
Kim, Dae Jung, 173
Kim, Jong Il, 175, 176
Kim, Yong Num, 175
Koizumi, Junichiro, 174, 176, 177, 200
Korean Peninsula, 35, 158, 159, 165, 170, 172, 173, 178, 181, 182, 196, 207, 214
Kosygin, Alexei, 56
Krepon, Michael, 194
Kuwait, 62

L

Landrieu, Mary, 133, 137
LDP (Liberal Democratic Party), 67
Libya, 19, 171, 217
Lieberman, Joseph, 133, 138
Lockheed-Martin Corporation, 46
Lowi, Theodore, 106

LSI (Lead System Integrator), 70, 73
Lyles, L. Lester, 69

M

Madison, James, 40
Material Interests, 83, 84
Mayhew, R. David, 105, 106
McHugh, John, 148
MD (Missile Defense), 25, 39, 48, 53, 76, 89, 90, 91, 92, 114, 118, 130, 152, 153, 157, 170, 195,198, 200, 201, 204, 214, 217
MDA (Missile Defense Act of 1991), 63
MDA (Missile Defense Agency), 22, 30, 45, 47, 49, 76, 77, 90, 94, 119, 153, 214, 226
MEADS (Medium Extended Area Defense System), 66, 140, 143
Meehan, Martin, 148
Military Research and Development Sub committee, 146, 148
MOU (Memorandum of Understanding), 68, 139
Mulvenon, James, 184

N

NASA (National Aeronautics and Space Administration), 59
Neustadt, E. Richard, 31, 102, 104
Nixon Administration, 80
Nixon, Richard, 57, 82, 84, 89
NAD (Navy Area Defense), 66, 140, 141, 142
NIE (National Intelligence Estimate), 70, 73
NMD (National Missile Defense), 15, 16, 20, 22, 31, 39, 41, 69, 70, 72, 73, 74, 75, 81, 83, 86, 87, 97, 98, 99, 102, 119, 120, 121, 125, 127, 128, 129, 131, 132, 134, 135, 138, 139, 141, 142, 143, 151, 152, 153, 158, 185, 190, 193, 216, 217
NMDA (National Missile Defense Act), 33, 72, 74, 92, 101, 130, 133, 134, 137, 138,153, 218
No Dong Missile, 117, 144
North Dakota, 120
North Korea, 18, 19, 36, 117, 126, 127, 144, 159, 160, 164, 165, 166, 170, 171, 172, 173, 174, 175, 176, 177, 178, 179, 180, 194, 217, 219, 220
North, Robert, 169
NPT (Nuclear Non-Proliferation Treaty), 159, 164, 172, 176
NSC (National Security Council), 50, 224
NTW (Navy Theater-Wide), 66, 67, 140, 142

O

OMB (White House Office of Management and Budget), 47

P

PAC-3 (Patriot Advanced Capability-3), 66, 140, 141, 142, 143, 146, 204, 205, 206
Pacific Ocean, 67, 206
Paek, Nam Sun, 180
Pakistan, 125, 131, 160
Panama, 32

Pareto, Vilfredo, 40
Persian Gulf War, 19, 62, 63, 91, 109, 125, 151, 202, 204
Pickett, Owen, 148
Pluralistic Security Community, 160, 206, 222
Policy community, 30, 42, 50, 51, 52
Position taking, 105
Powell, Colin, 180, 199
Post-Cold War, 19, 91, 98, 157
Pyongyang government, 165

R

Raytheon Corporation, 46, 63, 91
Reagan Administration, 60, 85, 90, 110, 111, 116
Reagan, Ronald, 18, 29, 30, 41, 59, 60, 64, 80, 82, 84, 86, 108, 137, 150
Republican Party, 68, 109, 118, 141
Reyes, Silvestre, 146, 148
Riley, Bob, 148
Riyadh, 202
Rogue nations (states), 15, 19, 98, 151, 171, 174, 217, 221
Roh, My Hyun, 177
Rosecrance, Richard, 170
Rumsfeld Commission (Report), 70, 71, 73, 92, 101, 126, 127, 128, 129, 217
Rumsfeld, Donald, 125, 205
Russett, Bruce, and Huth, Paul, 163, 166
Russia, 174, 178, 179, 220

S

SALT I (Strategic Arms Limitation Talks), 58
SALT II (Strategic Arms Limitation Talks II), 108, 131
Sanchez, Loretta, 148
Saudi Arabia, 62, 144
Scarborough, Joe, 148
Schultz, George, 59, 60
Scud missile, 202
Select Intelligence Committee, 133
Senate, 77, 92, 103, 106, 109, 115, 121, 122, 123, 130, 131, 136, 137, 138
Senate Armed Services Committee, 45, 46,
Sessions, Jeff, 128
SDI (Strategic Defense Initiative: Star Wars), 29, 31, 41, 59, 60, 64, 65, 66, 80, 81, 84, 85, 90, 91, 150, 170
SDIO (Strategic Defense Initiative Organization), 22, 29, 31, 60, 61, 66, 80, 83, 84, 86, 90, 91, 109, 214
SDS (Strategic Defense System), 61,
Shahab-3, 144
Shahab-4, 144
Six-nation talks, 179,
South Korea, 100, 120, 164, 165, 166, 173, 177, 178, 179, 193, 220
Soviet Union (USSR), 54, 58, 60, 91, 108, 114, 127, 150, 157, 161
State Department, 50, 70, 224
Straits of Taiwan, 24, 37, 158, 159, 188, 189, 197, 207
Strategic competitor, 185

T

Taepo-Dong 1, 67, 171
Taepo-Dong 2, 127
Taiwan, 35, 140, 162, 182, 183, 184, 185, 186, 187, 188, 189, 193, 197
Tam, Andrew, 188
THAAD (Theater High Altitude Area Defense), 66, 140, 142, 143, 153
The Philippines, 164
TMD (Theater Missile Defense), 23, 41, 64, 66, 67, 68, 69, 72, 73, 81, 83, 85, 98, 99, 109, 139, 140, 141, 143, 144, 146, 147, 153, 183, 216, 218, 219
TMDIA (Theater Missile Defense Improvement Act of 1998), 101, 139, 144, 147,
Triangulation, 130, 132
Turkey, 144
Turner, Jim, 148

U

UCS (Union of Concerned Scientists), 75
US Air force, 54, 89
US Army, 54, 55
US Missile Defense Policy, 41, 42, 52, 53, 98, 124, 125, 170, 185, 188, 189, 190, 191, 193, 204
UN (United Nations), 27, 196, 198, 220

V

Vandenberg Air Force, 74, 78
Vietnam, 32

W

Wang, Yi, 179, 180
War against Terror, 198
Weber, Max, 40
Weinberger, Casper, 60, 150
Welch, Larry, 74
Weldon, Curt, 130
Western Europe, 160, 206
White House, 19, 71, 115, 224
WMD (Weapons of Mass Destruction), 19, 151, 162
World Trade Center, 15, 20, 161, 199
World War II, 53, 86

Y

Yang, Dali, 185
York, Herbert, 57

About the Author

SETSUO TAKEDA is Professor of Political Science in the College of International Relations at Nihon University. He has written several books and numerous articles on U.S. foreign and security policy toward the Asia-Pacific region and U.S.– Japan relations.

2005年　6月30日　初版発行

著　者　　武 田 節 男
©2005 Setsuo Takeda

発行者　　高　橋　考

発行所　　三和書籍

〒112-0013　東京都文京区音羽2-2-2
電話　03-5395-4630　FAX 03-5395-4632
印刷・製本　　株式会社新灯印刷

乱丁、落丁本はお取り替えいたします。　価格はカバーに表示してあります。

ISBN4-916037-76-6　C3047